Gödel, Tarski and the Lure of Natural Language

Is mathematics "entangled" with its various formalisations? Or are the central concepts of mathematics largely insensitive to formalisation, or "formalism free"? What is the semantic point of view and how is it implemented in foundational practice? Does a given semantic framework always have an implicit syntax? Inspired by what she calls the "natural language moves" of Gödel and Tarski, Juliette Kennedy considers what roles the concepts of "entanglement" and "formalism freeness" play in a range of logical settings, from computability and set theory to model theory and second order logic to logicality, developing an entirely original philosophy of mathematics along the way. The treatment is historically, logically and set-theoretically rich, and topics such as naturalism and foundations receive their due, but now with a new twist.

JULIETTE KENNEDY is Associate Professor of Mathematics and Statistics at the University of Helsinki. Her research focuses on set theory, history of logic and philosophy of mathematics, and she is Editor of *Interpreting Gödel: Critical Essays* (Cambridge University Press, 2014).

Gödel, Tarski and the Lure
of Natural Language
Logical Entanglement, Formalism Freeness

JULIETTE KENNEDY
University of Helsinki

CAMBRIDGE
UNIVERSITY PRESS

University Printing House, Cambridge CB2 8BS, United Kingdom

One Liberty Plaza, 20th Floor, New York, NY 10006, USA

477 Williamstown Road, Port Melbourne, VIC 3207, Australia

314-321, 3rd Floor, Plot 3, Splendor Forum, Jasola District Centre, New Delhi - 110025, India

103 Penang Road, #05-06/07, Visioncrest Commercial, Singapore 238467

Cambridge University Press is part of the University of Cambridge.

It furthers the University's mission by disseminating knowledge in the pursuit of education, learning and research at the highest international levels of excellence.

www.cambridge.org
Information on this title: www.cambridge.org/9781108940573
DOI: 10.1017/9780511998393

First published 2021
First paperback edition 2022

A catalogue record for this publication is available from the British Library

ISBN 978-1-107-01257-8 Hardback
ISBN 978-1-108-94057-3 Paperback

For Jouko

Contents

Preface

This book is about *formalism freeness*, an idea stemming from Gödel's 1946 Princeton Bicentennial Lecture, or more precisely from the phrase "formalism independent" which appears in the opening paragraph of that lecture. The phrase "formalism independent" fell on fertile soil: The Helsinki Logic Group, my academic home for the last 20 years, is devoted to the semantic method in all its guises, whether it be the method of Ehrenfeucht–Fraïssé games, or Abstract Elementary Classes or other "logic-free" concepts, even as it is at the same time devoted to logic in its syntactic aspects. Secondly, the concept of "indifferentism", appearing in John Burgess's writings, solidified my interest in the constellation of ideas around formalism independence, formalism freeness, logic freeness, the pure semantic method and logics without syntax – ideas that pull away from foundationalism in ways that seemed puzzling to me at the time.

Gödel asks in his 1946 Lecture for notions of provability and definability that are formalism independent in the way he understands computability to be, given Turing's analysis of the intuitive notion: "human effective calculability following a fixed routine." Formalism independence in the case of computability involved confluence in the sense of transcendence with respect to a given class of formal systems. How to obtain confluence for definability? The following implementation suggests itself: given a notion of definability such as constructibility or on the other hand hereditary ordinal definability, the two notions Gödel considers in the lecture, why not vary the underlying logic with respect to these definability notions? Precisely: if, say, the constructible hierarchy is built over first order logic, then one can consider properties of the inner models (for set theory) one obtains from the constructible hierarchy, when built over other logics extending first order logic. In 2009 it was my great fortune to begin working with Menachem Magidor and Jouko Väänänen on exploring these inner models. This led to the paper [128], in which it was shown that

ix

new inner models can be obtained in this way, which are generically absolute and which contain certain large cardinals, among other results. This work is ongoing. Most of Section 4.4 is excerpted from [128].

My 2013 [124] explored the philosophical and historical aspects of the implementation offered in [128] as well as of what would eventually be called formalism freeness generally, a concept which was intended to broaden Gödel's notion of formalism independence. This book both builds on and greatly expands [124], albeit with a somewhat more conservative view of the model theorist's break with syntax in the background, than is evinced in [124]. In particular we distinguish the first order case from the case of strong logics, i.e. logics extending first order logic, which play a central role in this book. In the so-called Abstract Elementary Class context the break with syntax is definitive, but even in that case one can argue for the presence of an implicit logic. Passages in Sections 2.0, 2.1 and 4.2 are excerpted from [124].

Computability is the primary example of a formalism independent concept for Gödel. The book's Chapter 3, anthologising [123], explores the evolution of the concept of computability in the 1930s with an emphasis on Gödel's evolving views. I thank Alisa Bokulich and Juliet Floyd, the editors of the volume *Philosophical Explorations of the Legacy of Alan Turing,* for their permission to anthologise [123] here.

A word about methodology. We are inspired here by the *encyclopaedic novel.* And just as Herman Melville makes free to halt the narrative of *Moby Dick* for the sake of giving extended disquisitions on its various topical aspects, so we will often take licence to lapse into the informational mode in this book – not by inserting long treatises on ship-building, or on the biology of whales, but by enumerating, for example, the various semantic characterisations of first order logic; or by giving a slew of equivalents of weakly compact cardinals.

This book, long in the writing, owes a debt to many. Conversations essential to me during the writing of my 2013 [124] were had with John Baldwin, who recognised the interest of formalism freeness early on and then ran with it in his own book. Throughout the writing of this book Andres Villaveces and Norma Claudia Yunez provided essential input and perspective. I first spoke about formalism freeness in an ASL special session devoted to Gödel on the occasion of Gödel's centenary, to which session I was invited by Scott Weinstein. In addition to conversations over the years Scott read the manuscript carefully in fall of 2019 and made deep and interesting comments, also as to suggestions for work going forward. Andy Arana, whose clarity and creativity as a philosopher I have often benefitted from over the years, opened up the discussion with a number of observations, some of which made their way into this book. In the last year or so of writing, John Baldwin, Patricia Blanchette, John Burgess, Walter Dean, the late Mic Detlefsen, Sebastian Gandon, Tapani Hyttinen, Sandra Laugier, Penelope Maddy, Ofra Magidor, Maryanthe Malliaris, Juliet Mitchell, Gil Sagi, Zeynep Soysal, Silvia de Toffoli, Boban Velickovic, Dag Westerståhl and Mark Wilson encouraged, listened and in some cases contributed to the final shaping of the ideas. For pressing me on various weak points I am indebted to Anand Pillay for a long conversation we had one rainy night in New York City's financial district. In connection with the writing of the computability paper anthologised here I am very grateful to Wilfried Sieg, who made important comments and corrections to an early draft, and to Juliet Floyd for helpful comments, for her support of this project, and for essential conversations and friendship over the years. I would also like to express my gratitude to my editor Hilary Gaskin, for her Job-like patience, also to the friends and family who have supported me throughout the writing of this book, of which I mention Robert Disch, Kathrin Hilten, Roope Rissanen and Amy Sandback.

I am privileged to be surrounded by an exceptional community of logicians and set theorists. For memorable conversations and correspondence about the foundations of mathematics, stretching over decades and sometimes under extreme weather conditions, I would especially like to thank Joan Bagaria, Mirna Džamonja, Wilfrid Hodges, Saharon Shelah, John Steel, Boban Velickovic, Philip Welch and Hugh Woodin.

Much of this book was finalised in the fall of 2019 when I was a visitor at the Institute for Advanced Study in Princeton. I wish to take this opportunity to express my gratitude to Peter Goddard of the IAS for his support of my visits to the IAS over the years, and also to the School of Historical Studies at the IAS for providing such a supportive and warm scholarly environment during the years 2011–12 and subsequently. I am also grateful to the librarians managing the Gödel archive at the IAS, especially Marcia Tucker, for their help in archival matters and also for granting permission to quote from Gödel's unpublished Max Phil notebooks.

This book was finished while I was a guest of the Hebrew University in the late fall of 2019. The logical and set-theoretical atmosphere of the seminars at the Hebrew University is like that of no other, and it has been a great privilege for me to take part in them through numerous visits in recent years.

As one of many who have been the object of his generosity, both intellectual and personal, I would like to express my deepest gratitude to Menachem Magidor. It would offend Menachem's modesty to express my admiration for him in the terms I would like to do here. Suffice it to say that for me he personifies everything that is good – both in academic life and outside of it, as a human being.

Finally I would like to thank the dedicatee of this book, my husband, Jouko Väänänen. His vast knowledge of logic and set theory, his logical morality – the "border-crossing point of view" as I called it in a paper for his 60th birthday volume – and finally his sense for the depths in foundational practice, has profoundly shaped my own thinking. Our conversations about strong logics in particular were especially important for the development of the ideas of this book, and indeed, more than any of the few whose work and whose conversations with me are cited here, it is his voice which is heard in these pages – and how could it be otherwise? Such is love.

1

Introduction

Logic ruptured at its core in the twentieth century, when concerns about the consistency of mathematics led to the emergence of *foundational formal systems*, lighting the spark that split logical space into its syntactic and semantic modes. Of course the concept of a formal semantics came along fairly late – a fact that reveals much about the nature of foundational practice, as we will note repeatedly in what follows. It must also be said that the syntax/semantics distinction had been anticipated by many in the nineteenth century, most notably by Bolzano.[1] What is of interest here is the consequent inheritance of this rupture, namely a tripartite foundational terrain consisting of syntax, semantics and natural language, the latter being a kind of moving target caught somewhere between the previous two – exactly where it is caught, or whether it is caught at all, depending upon one's point of view.

The rupture was not seamless, but ragged – which is just to say, the interactions of syntax, semantics and natural language that played out across this foundational terrain, turned out to be complex. Morphologies of natural language would be refracted through the lens of the syntax/semantics distinction; and questions of the adequacy of our formal languages remain unresolved. The adequacy problem is a serious one: the foundationalist's view of mathematics' being enclosed by a formal language, of being disclosed and supported by that language, even as it is distorted and disrupted by it, is a fundamental commitment.

We will recount key points in the history of the syntax/semantics distinction below. Our main interest in this book is what had slowly begun to dawn on the logicians of the early twentieth century, namely the phenomenon of

---

[1] Some writers trace the distinction back to Aristotle. See Curtis Franks's Helsinki Lectures of 2015, www.helsinki.fi/sls2015/materials/Franks.pdf. See also T. Smiley's "Aristotle's Completeness Proof" [239].

entanglement: the fact that certain canonical mathematical objects are remarkably sensitive to slight perturbations of syntax and logic; and on the other hand the dual phenomenon of *formalism independence*, the idea that certain canonical concepts and constructions are stable across a variety of conceptually distinct formalisations, so seemingly insensitive to perturbations of logic and syntax.

For entanglement, consider the axiomatisation of the natural numbers due to Peano. Construed as a first order system the Peano axioms have continuum many countable models. Formulated as a second order system, those axioms have a unique model. For an example from set theory consider the Levy Reflection Principle, which is provable in ZF set theory, construed as a first order theory.[2] However if second order parameters are allowed in the reflecting formula, one obtains a strong axiom of infinity, which is unprovable.[3]

For another example, Leslie Tharp's result [255] that monadic third order logic interprets full non-monadic second order logic may be contrasted with the intuition that the passage from second to third order logic is minor in effect, at least such is true in the setting of the hereditary ordinal definable sets.[4]

These examples are a matter of logic but just considering signature,[5] languages that omit function symbols, for example, behave differently on the semantic side than languages that include them. A case in point are the zero–one laws for finite structures, which are sensitive to signature in the following sense: the probability that a random relational structure on the domain $\{1, \dots, n\}$ satisfies a given first order formula tends to either 0 or 1 as n tends to infinity. But if we allow function symbols as part of the language, even the simple sentence $\exists x (f(x) = x)$ has limit probability $1/e$.[6] Or consider the

---

[2]   The Levy Reflection theorem says that for any sentence ϕ of set theory we can prove in ZFC the implication $\phi \to \exists \alpha (V_\alpha \models \phi)$. Intuitively: Any sentence that is true in V, the cumulative hierarchy of sets, is true in some level V_α, i.e. anything that is true is true in a set size universe. In its stronger form the theorem says: Given n we can construct a (definition of a) club class C of cardinals κ such that $V_\kappa \prec_n V$. Alternatively, $H(\kappa) \prec_n V$. Reflection is a very general phenomenon in set theory.

[3]   Azriel Lévy, "Axiom Schemata of Strong Infinity in Axiomatic Set Theory," [151]. Hauser and Woodin observe that the move from first to second order parameters in this case is unproblematic: "From a heuristic point of view, the inclusion of second order parameters in the reflecting formulas is unproblematic because, for one thing, such parameters possess a canonical interpretation in the structures to which one reflects . . . Moreover, that interpretation extends the way in which definable proper classes may be incorporated in the proof of reflection in ZF." [98], p. 401.

[4]   See also [178].

[5]   i.e. syntax.

[6]   Result due to R. Fagin, see [56]. Note that relation symbols and function symbols are definable from each other.

Bernays–Schoenfinkel–Ramsey class,[7] which is decidable if we do not have function symbols. Whereas Y. Gurevich proved in [95] that if we have function symbols the class is undecidable, in fact even just the universal theory is.

For an example of sensitivity to syntactic resources, consider the real numbers. Conceived of as the complete ordered field with signature $\langle +, \times, <, 0, 1 \rangle$, their theory is decidable by Tarski's Theorem.[8] If one adds to the signature a symbol for the integers, or just the sine function, decidability fails.[9]

As for the dual notion of formalism independence, this was the phrase Gödel applied to the concept of computability in his 1946 Princeton Bicentennial Lecture [86]. *Confluence* was the term used by Robin Gandy in [77] to describe the fact that Gödel was gesturing at by describing the concept of computability as "formalism independent", namely that the class of recursive functions are stable across a wide variety of different *formalisms*. That is, whether one defines the notion of computability by means of the Gödel–Herbrand–Kleene definition (1936), Church's λ-definable functions (1936), Gödel–Kleene μ-recursive functions (1936), Turing Machines (1936), Post (1943) systems or Markov (1951) algorithms,[10] to mention just a few of these systems, one ends up with the same class of functions.

Another occasion on which Gödel mentioned the concept of formalism independence, among the many that appear in his writings, was in his 1940 Brown University Lecture, when he described the consistency proof for the continuum hypothesis as "absolute, i.e. independent of the particular formal system which we use for mathematics". Gödel went on to explain that the proof goes through for systems of arbitrarily high type; but he also left the door open for a wider interpretation of the term "absolute", so relative to a wider class of formalisms than those appearing in a particular type hierarchy.[11]

For a modern example from set theory it is known that certain canonical set-theoretical structures can be constructed over a large class of very different *logics*, in the sense that whether one uses this or that logic in the construction, one obtains the same structure.[12] How to measure the difference between

---

[7] The class is defined as the set of sentences that, when written in prenex normal form have a quantifier prefix of the form $\exists^* \forall^*$ and do not contain any function symbols. Here $*$ means you can have a finite string of similar quantifiers. See [19].

[8] i.e. its first order theory is decidable.

[9] The real numbers are in another sense independent of their construction, a point we will take up below.

[10] See, e.g. [45].

[11] [*1940a] in [87], p. 184.

[12] The structures referred to here are the constructible hierarchy L, and the class of hereditarily ordinal definable sets HOD. See Section 4.4.

logics? The Lindström characterisation of first order logic provides one metric, in the sense that the closer the logic comes to satisfying the Lindström conditions, the closer it is, we want to say, to first order logic. But we will consider other "metrics" below.[13]

In model theory formalism independence, or what we call in this book *formalism freeness*, can show up in the form of theorems giving semantic equivalents of syntactic concepts. For example in so-called *o-minimal* structures, a central area of research nowadays, the first order definable sets are just finite unions of open (or half open) intervals and their complements. This has the effect of translating a fundamentally syntactic concept, namely definability, into the language of "interval." Real closed fields, of which the field of reals are an example, are all o-minimal, by a result due to Tarski in 1929, as is the field of real numbers with exponentiation as an added function.[14]

In fact for some time now, in the wake of Tarski's work, the tendency to work in a logic-free setting has emerged in certain specialised areas of model theory. This is seen in the desire to dispense with metamathematical tools such as compactness; to dispense with the notion, even, of formula.[15] We will examine this tendency in depth in this book, from the perspective of both historical and contemporary model-theoretic practice. As we will argue in Chapter 5, a certain formalism free paradigm played a fundamental role in Tarski's logical work, taking the form of an explicit programme to replace metamathematical results by their, in Tarski's words, "mathematical" analogues. The paradigm was inaugurated by Tarski's 1929 result on definable sets of real numbers – *normalising* definability for the mathematician.

As for the work of model theorists subsequently, working in Tarski's tradition, we will argue that while the tendency has been toward working *purely* semantically, which we define here as working entirely in natural language, at the same a methodology has emerged based on the dynamic, episodic and *transient* use of syntactic methods, a kind of "on again off again" use of syntax and logic, in which adequacy constraints play no role.

Entanglement and formalism independence occur everywhere in foundational practice, viewing foundational terrain through that particular lens. The

---

[13] See Section 2.2.2 for a definition of the Lindström conditions. See also [153].

[14] By Wilkie's theorem [283]. Of course, the robustness of the concept of o-minimality has to do not with formalism independence of the set-theoretic definition, but rather with yielding consequences for many theories that are apparently only distantly related.

[15] In *abstract model theory* for example, one replaces the syntactic notion of sentence by the class of its models.

analysis of these two concepts can be systematised in various ways. In Section 4.4 we will present a calculus based on the *parametric treatment of logics*, defined as follows: consider a canonical mathematical object or construction, given in natural language. By formalising the construction we mean reconstructing the object within a formal system, one equipped with a specified formal language, an exact proof concept and an exact (formal) semantics, such that the proof concept is sound with respect to the associated semantics. One can ask whether the natural language object is (when formalised) particularly sensitive to aspects of the underlying logic or formalism, in the sense that a slight change in the formalism, e.g. on the level of syntax, leads to a significant change in the formal environment.

If there is such a significant effect, then in this case we say that the object is *entangled* with the formal systems in question. If the object remains the same in spite of being built over a large class of different formalisms, where the difference is measured by, for example, deviation from the Lindström characterisation of first order logic, then we would say that the object exhibits a degree of *formalism freeness*. We will develop this calculus in chapter 4, in which we aim to implement Gödel's 1946 suggestion to develop notions of provability and definability which are language independent in the way Gödel understood computability to be. Specifically, the calculus we will develop there is based on treating logics parametrically in the setting of definability in set theory. In Chapter 6 we will offer a second calculus, based on the concept of *symbiosis*, aimed at studying the set-theoretical entanglements of a logic.[16]

A caveat: we used the phrase "slight change in the formal environment" above, but of course the change from a relational to a functional language, or from first to second order quantification, or any other move one might make in the way of adjusting one's syntax or logic in setting up a formal framework for mathematics, is not a slight thing at all to the logician – just as the flapping of a butterfly's wings is no slight thing to the butterfly, or indeed to the chaos theorist measuring catastrophic outcomes.

Of course from the perspective of so-called "core" or classical mathematics, the kinds of logical moves we are discussing, the choices of logic and syntax that are essential in the framing of a foundational formal system, are simply missing from the natural language practice of the mathematician. In that sense one may consider mathematics itself, the mathematics of the everyday working mathematician, to be entirely formalism free. A ready to hand illustration of

---

[16] In his 2004 monograph *The Collapse of the Fact/Value Dichotomy* [201] H. Putnam uses the word "entanglement" differently than here, arguing for the entanglement of valuation and factual description.

this can be found in the ordered field of real numbers, in that whether one builds the reals as equivalence classes of Cauchy sequences (built over, say, first order logic), or as Dedekind cuts (built over, say, second order logic), or in some other way, it is remarkable how impervious mathematical practice is to the set-theoretic definition of the reals.

John Burgess calls this phenomenon "indifferentism" in his [31] and [32], which he defines as the "general phenomenon of the indifference of working mathematicians to certain kinds of decisions that have to be made in any codification of mathematics . . . two analysts who wish to collaborate do not need to check whether they were taught the same definition of 'real number'". "Translation hiding in plain sight", as E. Apter calls indifferentism,[17] Burgess traces its advent to the mid-nineteenth century, when mathematicians established the arithmetisation of analysis, freeing it from its geometrical foundations. The mathematician is indifferent to logical foundations, and this is as it should be:

A "foundation" that required one to be constantly taking note of it would, it seems, be much less desirable than one that can in practice be allowed to repose largely unread in chapter 1 of a virtual encyclopedia now up to chapter 1001.[18]

Incidentally, Kreisel makes a related observation in his 1967 [141]:

. . . for, in his own work he [i.e. the mathematician JK] never gives a second thought to the form of the predicate used in the comprehension axiom! (This is the reason why, e.g., Bourbaki is extremely careful to isolate the assumptions of a mathematical theorem, but never the axioms of set theory implicit in a particular deduction, e.g. what instances of the comprehension axiom are used.)[19]

Foundational issues may lurk at the edges of subjects like analysis of course, often having to do with the axiom of choice, which is always problematic from the foundational point of view; or related to this the idea of an arbitrary subset of natural numbers. But these foundational issues are of a somewhat different ilk than questions concerning the choice of, say, signature.

For Burgess indifference to "actual codification" may kick in at a higher level (than that of logic); it is "a principle of never looking inside previous results", implying that

---

[17] "What Is Mel Bochner Translating? Some Thoughts on New Knowledge Alphabets and Inter-phenomenal Modes of Perception," lecture at *Colloque Mel Bochner. On translation*, ENS Paris-Saclay, December 2019.

[18] [32], p. 125.

[19] [141], p. 151.

if there are different routes by which the material (notions and results) on which the author immediately relies could have been obtained from first principles (primitives and postulates), the author can be indifferent to any choice among them.[20]

Of course there are many occasions in the natural language practice when issues of signature, in the logician's terminology, or in mathematical terms the choice of *primitive terms*, is relevant: in representation theory (in mathematics), for example, a single object can be considered from the point of view of different theories, each equipped with their own primitive concepts, theories such as modules, differential geometry and so on. But these are neither formal nor foundational theories, which is our concern here.

Entanglement and formalism independence are interesting phenomena in themselves – in fact much of the logician's daily work involves tracking the effects of syntactic variation. We will explore some of the technical aspects of formalism independence below. But there is also a philosophical moral to be drawn. Entanglement signals a breakdown in the adequacy of our formalisations by exposing misalignments between entangled formalisations and the intuitive concepts they are meant to capture. Speaking very broadly, any formal system already fails to be adequate in this sense, in that ordinary mathematical discourse very naturally employs quantification over objects of different orders, and is already wholly impervious to details of signature such as the difference between relational and functional languages. For example the completeness axiom of the order of the reals – every bounded non-empty set of reals has a supremum – is a second order axiom, while the other axioms of the reals are first order.

So any formalism, in that it must *by definition* be committed to a choice of quantification of a particular order, or committed to a choice of a particular signature, already fails to adequately capture the practice, at least on the level of surface grammar.

Formalism independence, on the other hand, has served as a sign of adequacy, even perhaps as an indicator of stable semantic content. Consider again the concept of computability. The formalism independence of the concept was crucial in bringing to rest – at least for a time – the debate among the logicians of the 1930s over whether the intuitive notion of "human calculability following a fixed routine" had been, so to say, "captured".[21]

---

[20] ibid, p. 125.

[21] We unpack the history of computability from the formalism independence point of view in Chapter 3.

Outside of foundations, that is within mathematics proper, there is of course a powerful tendency to resist context relativity, manifested in the valorisation of the idea of *robustness*, a topic which would easily deserve a book of its own. "Things are robust if they are accessible (detectable, measurable, derivable, definable, producable, or the like) in a variety of independent ways", in the words of William Wimsatt.[22] The question why and how robustness functions as a methodological ideal in mathematics is a complex one. We are concerned here though with the question how robustness in the form that Gödel identified it, i.e. formalism independence, drove the foundational practice of Gödel and Tarski, and continues to drive the work of contemporary logicians.

We will see that just as in the historical cases, following the trail of absoluteness, confluence, formalism independence and the like to its end, namely into natural language; seeing exactly where in foundational practice we are pitched into natural language, and why, permits the philosophical moves we want to make come clearly into view – laying bare, in the process, our own philosophical commitments.

1.1 The Syntax/Semantics Distinction

A second philosophical moral to be drawn involves the syntax/semantics distinction itself – a distinction we problematise, albeit tacitly, in this book. It is a distinction on which the categories of formalism freeness and entanglement supervene, and thus it is incumbent upon us to address the question, how sharp is the syntax/semantics distinction? How well-defined? There is clearly an imbalance between, on the one hand *syntax*, the notion of a finite string of symbols being an apparently clear concept; and on the other hand *semantics*, a term with many meanings in the literature. There are the semantics of formal languages as given, for example, by Tarski, Kripke (for modal logic), and others; there is semantics in the sense of a meaning theory, for example in natural language; there is the idea of semantic content, or, simply, of content, or meaning; there is the phrase "pure semantics" in the sense employed above, to mean "formulated in natural language".

---

[22] See [286]. See also Putnam's remark in "Mathematics without Foundations" [199]: "In my view the chief characteristic of mathematical propositions is the very wide variety of equivalent formulations that they possess." This specialises to foundations, according to Putnam, as witnessed by the equivalence of set-theoretic formulations on the one hand, with modal formulations on the other, of said propositions.

As for the sharpness of the distinction – the idea that syntax and semantics are separate and autonomous fields – this is challenged by the fact that semantic frameworks can be viewed as having an implicit syntax, even an implicit logic, as we will argue in chapter 6. There is also the fact that most semantic concepts can, with only a slight shift of attention, be converted into syntactic ones – thinking, for example, of formal satisfaction predicates.[23] And the converse is also true, as witnessed by the examples given in this book – syntactic concepts can easily be seen to admit robust semantic characterisations.[24] A semantic characterisation seems to convert the syntactic object into a kind of *ersatz*, and vice versa the conversion of a given semantic framework into a syntactic one may be thought of as absorbing the former into the latter.

In the hands of model theorists the syntax/semantics distinction is plastic. We mentioned Tarski's theorem characterising definable set of reals in topological/geometric terms. In the work of the contemporary model theorist B. Zilber, the syntax/semantics distinction is translated into the language of algebraic varieties (the syntax) and curves (the semantics).[25]

For J. Baldwin, the property of having countably many types over the empty set is syntactic, a property "more artfully phrased by saying the stability theoretic properties are 'properties of syntax' that are absolute".[26] Thus the property of a theory being ω-stable is syntactic because the condition asserts the nonexistence of an infinite branch in a certain finitely branching tree, which is absolute. This is in contrast to Vaught, writing in [272] that "a little thought convinces one that a notion of 'purely syntactical condition' wide enough to include [the property of having countably many types over the empty set JK] would be so broad as to be pointless".[27]

The model theorist A. Pillay highlights the fluidity of the distinction:

> The traditional notions of semantics and syntax are also fluid. Categorical logic gives an account of the basic syntactic object of first order logic, namely a complete theory, as a suitable category (even topos). Another point of view is to take the type spaces as the fundamental objects, so a first order theory is a suitable functor from the category of natural numbers to profinite spaces. From this point of view some of the syntactic aspects of model theory appears in functional analysis,

---

[23] Though we stop well short of the proof theorist's motto "semantics is just syntax in disguise".

[24] See especially Sections 2.2.2 and 5.1.1.

[25] "Around Logical Perfection" [112], J. A. Cruz Morales, A. Villaveces, and B. Zilber, to appear.

[26] Baldwin, personal communication.

[27] And Vaught did not even know the tree characterisation of ω-stability.

in other languages. So the old notion of semantics being about mathematics and syntax about logical formalism is modified. The formalism is itself an object of mathematics.[28]

It must be said that the syntax/semantics distinction is taken for granted, mostly, in foundational practice. This is in contrast to the philosophy of language literature, in which the syntax/semantics distinction is contested. So-called generative semanticists challenged the distinction;[29] and the category mistake – statements of the kind "Green ideas sleep furiously", to use a famous example of Chomsky – can be viewed as throwing the distinction into question.[30]

"Category mistakes can be meaningful", O. Magidor has argued, for otherwise assumptions about the functioning of language, e.g. as related to compositionality, can be called into question:

> What makes category mistakes particularly interesting is that a plausible case can be (and indeed has been) made for explaining the phenomenon in terms of each of syntax, semantics, and pragmatics . . . Consider for example the question of whether or not category mistakes are meaningful. If they are, one might be able to accept a strong form of the principle of compositionality, according to which any meaningful expressions combined in a syntactically well-formed manner, compose a meaningful expression. On the other hand, if category mistakes are not meaningful then, assuming they are syntactically well-formed, one can at best accept only a weaker principle of compositionality . . . [31]

The debates in philosophy of language that centre around syntax and semantics differ in important ways from the foundational debates in philosophy of mathematics. Nevertheless the questions that Magidor frames here resonate profoundly with foundational practice. Can syntactic features be read off semantic ones? How to classify mathematical concepts as either syntactic or semantic at all? And finally, is syntactic well-formedness sufficient for meaning? Magidor addresses this issue in connection with category mistakes:

---

[28] Pillay, personal communication.

[29] See [164], p. 18.

[30] As O. Magidor puts it:

In the 1960s and 70s the question of how to analyse category mistakes thus played an important role in the foundation of linguistics: because it was not clear whether the phenomenon should be treated as syntactic or semantic, the question of how to handle it formed a fruitful ground for exploring what the border between the two fields is, *if there is one at all.*

[164], p. 19. Emphasis ours.

[31] [164], p. 6.

Relatedly, since category mistakes are arguably the only example of sentences that might be considered syntactically well-formed but meaningless, the issue of whether they are in fact meaningless may decide a crucial question concerning the relationship between syntax and semantics: whether being syntactically well-formed is a sufficient condition for meaningfulness. Conversely, other foundational questions have a direct impact on one's theory of category mistakes. For example, as we shall see, *the issue of whether syntactic features may supervene on semantic features*;. . . Finally, exploring the issue of category mistakes can shed light on the fundamental methodological question of how one might decide whether to treat a phenomenon as syntactic, semantic, or pragmatic.[32]

In this book natural (mathematical) language is simply taken to mean "not formal", so in contrast to any analysis coming from the philosophy of language literature, the workings of natural language are left unanalysed by us, *insofar as those workings turn on specific usage*. Tacitly, so insofar as natural language concepts are considered in their state of being entangled with various formalisms, we *do* analyse usage.

The question above, whether syntactic well-formedness is sufficient for meaning, could well be posed in the foundational context, albeit with a different notion of meaning at stake. As for the question whether syntactic features supervene or in some sense can be read off semantic ones, the question is posed here, but localised to foundational (or metamathematical) practice. In particular we ask the question, whether and under what conditions a given model class, i.e. a class of structures closed under isomorphism, has a natural or implicit logic; or even a natural syntax. There is also the question under what circumstances a game may give rise to a logic.[33]

Our emphasis throughout is on the adequacy of formal systems with respect to natural *mathematical* language. The broader question of the validity of the key assumption in "Models and reality" [200] for example, namely whether the natural language of science altogether can be captured by a formalism, either in its meaning aspect or in its other, more directly formal aspects, will not be treated here.[34]

---

[32] [164], ibid. Emphasis ours.

[33] Games in the sense of logic. See Sections 2.2.3, 5.6 and 6.3.1.

[34] M. Stockhof's "Can natural language be captured in a formal system?" [244] finds the community to be divided on the question of his title. Those answering the question in the affirmative include Montague, Davidson, Lewis and Cresswell, according to Stockhof. On the no side Stockhof includes radical contextualists on meaning such as Charles Travis, along with ordinary language philosophers and others citing the inherent and intransigent vagueness of natural language.

1.2 Our Logical Pluralism

Prima facie, our approach assumes a pluralism of logics, not so much in the Carnapian sense as in the sense of M. Wilson's 2008 opus *Wandering Significance*, which emphasises the implicit *multiscalar* nature of scientific practice. As Wilson remarks, "This evaluation of our computational predicament is not pluralist in the usual philosopher's sense; it merely supplies straightforward scientific explanations of the diverse descriptive demands that make more impatient thinkers presume that we should become pluralists or anti-realists."[35]

For Wilson the necessity of multiscalar descriptive strategies, "the key to adaptive success, is to find ways to re-engineer a familiar reasoning scheme A into a fresh routine that addresses task B in a swift and relatively painless manner".[36] Wilson opens his monograph [285] with a concrete example: a bank robber on the way to the scene of the crime (or crime-to-be), in need of precise directions. The bank robber divides the task into distinct investigative stages: first using a large scale map of the state, then passing to a set of indications localised to the town, then finally to instructions of the form "the bank is between X and Y streets at the corner of street Z".

Wilson provides us with a framework for viewing the multiscalar, or as we will call it *prismatic* nature, not of applied mathematics but of foundational practice, insofar as that practice is concerned with adequacy; that is, driven not by concrete or empirical problems, but by the inadequacies inherent to various kinds of formal systems. For example, in a first order system such as Peano arithmetic one has a finitistically viable proof predicate (recursively enumerable), which seems to capture the common notion of proof. But first order logic imports referential opacity, e.g. in the case of first order arithmetic theories there are continuumly many interpretations, taking just the countable models, as we noted above. In the second order formulation of arithmetic referential opacity disappears in the sense that we have a unique model, but then we lose our finitistic notion of proof, in fact the complexity of the second order proof concept is Π_2-complete in the Levy hierarchy.[37]

A prismatic foundation, then, made up of a patchwork of theories. What then of theoretical unification, the problem of what occurs at the border between theories? In this book we suspend what one may call the "foundational attitude", namely the idea that Wittgenstein was so critical of in his later writings, that "at the deepest level of language there is an underlying logical structure

---

[35] [284], p. xvi.
[36] [285], p. xiii.
[37] [255].

which runs through our thought and language";[38] the idea of essential structure; or as S. Laugier puts it, the idea that we always speak from within a (foundational) theory.[39]

For if mathematicians speak from within a theory, then one may ask, which theory? ZFC set theory is a convincing candidate, as is category theory. There is also higher order logic, and indeed many fundamental concepts from analysis and other areas of mathematics have a very natural second order definition, such as: induction over the natural numbers; the completeness of the ordering of the reals; the concept of a linear order being well-ordered; the concept of a structure being countable; the freeness of a group. In fact none of these are first order definable without building set theory around the concept in question.[40]

But can one really choose between, say, set theory and second order logic? J. Väänänen has argued that from the point of view of the practice mathematics altogether is indifferent to such a choice, especially when that choice is between first order set theory and higher order logic. His view is that the working mathematician will – and should – be indifferent to the choice, and there are deep theoretical reasons why this should be the case:

> We study two metatheories of mathematics: first order set theory and second order logic. It is often said, that second order logic is better than first-order set theory because it can in its full semantics axiomatize categorically $\mathbb{N}$ and $\mathbb{R}$, while first-order axiomatization of set theory admits non-standard, e.g., countable models. We show below that this difference is illusory. If second order logic is construed as our primitive logic, one cannot say whether it has full semantics or Henkin semantics, nor can we meaningfully say whether it axiomatizes categorically $\mathbb{N}$ and $\mathbb{R}$. So there is no difference between the two logics: first-order set theory is merely the result of extending second order logic to transfinitely high types.[41]

Putnam, made the same point in "Models and reality":

> The "intended" interpretation of the second-order formalism is not fixed by the use of the formalism (the formalism itself admits so-called "Henkin models", i.e., models in which the second-order variables fail to range over the full power set of the universe of individuals), and it becomes necessary to attribute to the mind special powers of "grasping second-order notions".[42]

---

[38] Cora Diamond, lecture, APA Eastern Division Meeting, January 2019.

[39] [148], p. 30.

[40] See [267] for a discussion of the foundational aspects of second order logic. See also Shapiro [218].

[41] See [264], pp. 506–507.

[42] [200], p. 481.

We will return to the inscrutability of semantics in the second order setting in Chapter 2.[43]

1.3 Formal vs Linguistic Semantics

A word about the relation between formal semantics and linguistic semantics, that is, semantics in the sense of a meaning theory for natural language. The first thing to note is that the natural language of the mathematician is, in the logician's terminology, an *interpreted* language. J. Baldwin puts the point this way:

> Consider the notion of a polynomial. In a standard high school Algebra I book ([Educational Development Center 2009]) a polynomial is defined (with no fanfare of syntax and semantics) as a sum of monomials where a monomial is earlier defined as a product of variables raised to non-negative integer powers and a (usually real) number coefficient. In this style of development a polynomial function is a map ... defined by a polynomial. The interpretation of the formal language in the structure that is fundamental to Tarski's definition of truth is made matter of factly in elementary algebra. Then operations on the ring of polynomials is smoothly defined (not named) in a concrete way by example and previous experience with 'combining terms'. In contrast, the definition of a ring of polynomials in Lang's Algebra book [Lang 1964] is much more abstract but avoids even a glimpse of the syntax-semantic distinction. There polynomial functions are defined by a composition of functions.[44]

Baldwin may be using a broader sense of "interpretation" than the logical one. Tarski, working in the context of a direct and uncritical relationship to natural language, certainly seems to use a broader sense of the term "interpretation" here:

> It remains perhaps to add that we are not interested here in 'formal' languages in sciences in one special sense of the word 'formal', namely sciences to the signs and expressions of which no material sense is attached. For such sciences the problem here discussed has no relevance, it is not even meaningful. We shall always ascribe quite concrete and, for us, intelligible meanings to the signs which occur in the language we shall consider. The expressions which we call sentences still remain sentences after the signs which occur in them have been translated into colloquial language. The sentences which are distinguished as axioms seem to us materially true, and in choosing rules of inference we are always guided by the principle that

---

[43] As for first order vs higher order theories in physics, while space–time theories admit a first order formalisation, it has been suggested that quantum mechanics admits only higher order formalisations. Zilber, personal communication, and Toader, unpublished manuscript. For the first order case see Andréka, H., Madarász, J.X., Németi, I. et al. "A logic road from special relativity to general relativity" [4].

[44] [10], p. 301.

when such rules are applied to true sentences the sentences obtained by their use should also be true.[45]

The view taken in this book is that precisely because the natural language of the mathematician is in practice an *interpreted* one, even in this vague sense of interpretation, that practice is tracked most closely on the semantic side of foundational work. In that sense semantics serves as a kind of synaptic connection laying down a pathway between two languages, to wit: formal and natural language. From this point of view it is fitting that Tarski's foundational programme involving semantic reformulations of metamathematical concepts was viewed by him as pulling syntax not just into semantics, but pulling syntax all the way into natural language.

Returning to the variegated meanings of the term "semantic": J. Burgess writes in "Tarski's Tort" of the confusion between *models* and *meaning*, of

> ... the evils of confusing, under the label "semantics," a formal or mathematical theory of models with a linguistic or philosophical theory of meaning. Tarski's infringing on the linguists' trademark "semantics," and transferring it from the theory of meaning to the theory of models, encourages such confusion, which has several potentially bad consequences.[46]

Burgess points out various confusions for quantified modal logic arising from, for example, extrapolating from a rigorous model theory, to a rigorous semantics, to a rigorous meaning.[47] Truth-conditional semantics

---

[45] [251], p. 166. See also Woleński, who traces the view to Leśniewski:

Having endeavoured to express my thoughts on various particular topics by representing them as a series of propositions meaningful in various deductive theories, and to derive one proposition from others in a way that would harmonize with the way I finally considered intuitively binding, I know no method more effective for acquainting the reader with my logical intuitions than the method of formalizing any deductive theory to be set forth. By no means do theories under the influence of such formalizations cease to consist of genuinely meaningful propositions which for me are intuitively valid.

Leśniewski [150], pp. 487–488.

[46] [31], p. 12.

[47] The quote continues:

Such a confusion may lead, on the one hand, to erroneous suspicions that many ordinary locutions involve covert existential assumptions about dubious entities. (For example, one may fall into a fallacy of equivocation and argue that since possible worlds are present in the model theory of modal logic, they are therefore present in the semantics of modality, and are therefore present in the meaning of modal locutions.) Such a confusion may lead, on the other hand, to unwarranted complacency about the meaningfulness of dubious notions. (For example, one may fall into a different fallacy of equivocation and argue that since quantified modal logic has a rigorous model theory, it therefore has a rigorous semantics, and therefore has a rigorous meaning.)

(which Burgess rejects) also gets an unwarranted boost, "through their mistaken association with the prestigious name of Tarski". And the confusion between models and meaning has consequences for the so-called Benaceraff's challenge:

> The tendency Bernays called "platonist" can lead to nominalist skepticism if combined with the identification or conflation of "formal semantics" or model theory with "linguistic semantics" or meaning theory.[48]

The confusion between models and meaning, the habit of conflating the two (associated notions of semantics), is precisely what we are interested in here – not for itself but for how it plays out in foundational practice; how it is resolved, and ultimately, how it lends power to, for example, the model theorist. The model theorist works semantically, or even purely semantically, in the sense of dealing with theories interpreted in natural language; but s/he will also rent space from formal languages – from syntax, on an as needed basis. The structure in question is met on two fronts: as a natural language object and as an object interpreting a formal theory. The model theorist lives in two linguistic worlds; what draws our attention is the fact that this divided existence has led to deep theorems – Mordell–Lang being one example, along with many others following that bifurcated schema.[49]

Baldwin cites Faltings theorem in connection with "context-changing domain changes", and the shift in the conclusion of the theorem from a purely group-theoretic statement to a statement about the stability of the relevant induced structure:

> The transformations required to state this conjecture represent several context-changing domain changes ... The problem is transformed from the original wild environment of the natural numbers to the slightly tamer rationals and then to the domestic algebraically closed field. Conceptually, analytic and topological tools came into play in defining genus and the Jacobian variety. But all of this work was carried out informally. Recent works shows the relevance of a full formalization. But the particular formalization depends on prior work in abstract stability theory that provides a fruitful framework.[50]

As for the reason why the mathematicians' semantic mode of thought feels so natural to him or her, this is weirdly verified by a variant of Gödel's speed-up theorem:

---

[48] Burgess, course notes for graduate course "Philosophy of Mathematics: Mathematical Ontology," PHI 536.

[49] We elaborate the point in Section 6.1.

[50] [10], p. 148.

Theorem 1.3.1 *There is no recursive function h such that the following holds:*
If ϕ is a first order sentence and there is a proof in ZFC *of $\forall M(M \models \phi)$ with*
n symbols, then there is a predicate calculus proof of ϕ of $\leq h(n)$ symbols.[51]

This can be interpreted as saying that the semantic method, the method of
establishing logical consequence through the use of models,[52] enjoys a so-
called "speed-up" over the method of formal proofs. As the authors write,

> We argue that the semantic method of proving logical consequence, based on
> Gödel's Completeness Theorem, is in an exact sense the most effective possible
> method. This is in harmony with the observation that the quickest way to prove a
> logical consequence $\phi \vdash \psi$ in practice is to take an arbitrary model of ϕ and show
> that it is a model of ψ.

In conclusion: we have on the one hand the natural language practice of
mathematics, a practice answerable to a variety of imperatives: methodological, aesthetic and programmatic, among others, while at the same time that
practice is impervious to perturbations of logic and syntax, or as we called it,
formalism free; and on the other hand we have the logical moves of the foundationalist seeking to adequately capture that practice, moves which are often
unstable with respect to language, when one sees stability so abundantly on the
side of the practice in its natural language mode. Two attractors pulling toward
each other: natural language mathematics in possible need of a grounding, and
logical practice in search of that ground.

The aim of this book is to activate the area between these two poles. To give
a granular account of historic and programmatic data we think of as relevant,
while at the same time prescinding from arguing on a more general level –
arguing in favour of, say, the semantic point of view. We do not wish to argue
against formalisation – for, to paraphrase E. Apter, if one is against formalisation, as a logician, what could one possibly be for?[53] Only our interest here is
in alternative uses.

Ours is not the logician as foundationalist, then, or as first philosopher.[54]
Ours is the logician as chaos theorist, tracking the impact on the "system", of
perturbations in the initial logical framework. We confine ourselves to a close

---

[51] M. Vardi and J. Väänänen, unpublished manuscript.

[52] So ψ is a consequence of ϕ if every model of ϕ is a model of ψ.

[53] From *Against World Literature: On the politics of untranslatability*:

> In titling this book, *Against World Literature: On the politics of untranslatability*, an
> interrogation shadows the provocation of the fore-title: if one is *against* the revival of World
> Literature in some of its new institutional guises, then what is one *for*?

[6], p. 2.

[54] First philosopher in the sense of e.g. P. Maddy's [159].

study of particular cases, offering "provisional progress reports", as M. Wilson refers to his own work[55] "with respect to how things presently appear." Our approach is historical, mathematical and in the Wilsonian sense pragmatic and philosophically prudent, aimed as it is at tracking our various investigative contexts; investigating the "cognitive architecture" of the situation, and developing on that basis various calculi aimed at measuring degrees of entanglement. Two calculi are offered: in Section 4.4 we ask, and attempt to answer, the question, how sensitive is a structure or concept to an underlying logic or formalism? In Section 6.5 we look for degrees of logical content in naive set-theoretic language.

[55] p. xvii.

2

Formalism Freeness and Entanglement: Definitions

When Gödel proved his Incompleteness Theorems he left open what an effectively given formal system means. Only after Turing's fully *mathematical* definition of effective computability was given, was Gödel ready to declare the concept of a formal system to be clearly defined: a formal system can be thought of as any mechanical procedure for producing formulas.[1] In this book we adopt a precise notion of *formalism*, namely: a syntax, a set of rules for building terms and formulas, a list of axioms, and finally a list of rules of proof. A formalism may be considered to include an associated semantics. As usual we assume that the proof concept of the formalism is sound with respect to the associated semantics.

In this book we will sometimes use the terms "logic" and "formalism" interchangeably. Whether we mean axiom systems, such as Peano Arithmetic and ZFC set theory, or formal languages such as first and second order logics is intended to be clear from the context. We define the notion of an abstract logic in Section 2.2.2.

Of course some question whether a formalism needs a semantics at all, or regard semantics as an afterthought, a possibly helpful but otherwise inessential addition to the logic. This is not the concept of formalism or logic that we use. In this book semantics – whether formal, e.g. model- or set-theoretic, or informal – is our main interest and the other aspects of a formalism are subordinate to it. By and large we assume that the various lists and rules that are constituent of a logic are effectively given. In such a case we call the formalism

---

[1] Gödel goes on to say that "For any formal system in this sense there exists one in the [usual] sense that has the same provable formulas (and likewise vice versa)..." [84], p. 369. We consider this and the surrounding remarks in depth in Chapter 3.

finitary, though we also consider infinitary formalisms, such as the infinitary languages $L_{\kappa\lambda}$.

With this concept of formalism and inspired by Gödel's use of the expression "formalism independence" we introduce the term *formalism freeness* to denote **the suppression of any or all of the above aspects of a logic or formalism, except semantics**. By introducing the term formalism freeness and including under its banner not only the concept of computability but in addition concepts that are stable with respect to permutation of the "underlying logic", the logic used to formalise the concept, as well as other forms of syntactic variation, we generalise Gödel's notion of formalism independence.[2]

Formalism freeness is view multi-faceted. We mentioned indifference to a choice of codification, or of syntax; but there are other forms of indifference, such as indifference to the choice of a semantics. We now take a moment to explore this aspect of formalism freeness.

It has been argued that mathematical discourse in its second order mode is indifferent to a choice between full and Henkin semantics. We argued for this in our [129], depending in turn on Väänänen's [266], in which it was argued that from the point of view of mathematical practice, when we actually use second order logic we do not and in fact cannot see a difference between ordinary ("full", or "standard") models and general or Henkin models:

> I will argue in this paper that if second order logic is used in formalizing or axiomatizing mathematics, the choice of semantics is irrelevant: it cannot meaningfully be asked whether one should use Henkin semantics or full semantics. This question arises only if we formalize second order logic after we have formalized basic mathematical concepts needed for semantics. *A choice between the Henkin second order logic and the full second order logic as a primary formalization of mathematics cannot be made*; they both come out the same.[3]

For example, let us consider Bolzano's Theorem:

Theorem 2.0.1 *(Bolzano) Every continuous real function on* [0, 1] *which has a negative value at* 0 *and a positive value at* 1 *assumes the value* 0 *at some point of* (0, 1).

---

[2] In the chapter of his [10] devoted to the mathematical properties of formalism freeness J. Baldwin takes a definition to be formalism free "if it is given semantically without any formal distinction between syntax and a semantics". Formalism *independence* is used by Baldwin to denote what Gandy called confluence, i.e. the equivalence of conceptually distinct definitions of a mathematical concept. [10], pp. 300–301. In this book we certainly include a lack of any formal distinction between syntax and a semantics under the banner of formalism freeness; but our concept is also broader. As for the phrase "formalism independence" we use the phrase here strictly to denote the various meanings Gödel attached to the phrase in his writings.

[3] [266], p. 505. Emphasis ours.

For the proof, by the second order comprehension axiom one can instantiate a universal second order quantifier at $X = \{x \mid f(x) < 0\}$. The set X is even first order definable, with f as parameter.

This is a paradigm example: we operate on sets definable from existing sets – *and these definable sets already exist in any general model*. Of course, principles such as the Axiom of Choice force us to introduce also non-definable sets, but they do not exist because "all" sets exists but because we assume – and the general models are assumed to satisfy – the Axiom of Choice. Because of the comprehension axiom, then, full semantics in some sense always "collapses" to Henkin semantics. From this point of view the distinction between the full and the Henkin semantics is blurred.

Returning to formalism freeness in general, as for the concept of a vocabulary, of course vocabulary in the *informal, natural language sense*, that is, as detached from any formalism based on it, is always a residue of the practice and in that sense is not suppressed in the applications we will consider here (or for that matter in mathematical investigations generally). The distinction we draw is between vocabulary thought of as the set of primitive terms of a mathematical area, such as group theory, and vocabulary as a constituent of a formal language. In (the informal version of) axiomatic set theory for example, what is emphasized is "the intuitive notion of the cumulative type structure," as Kreisel puts it in his [137], over the formal theory ZFC.

J. Baldwin explains the distinction in his [9]:

> It is in this sense that certain recent work of Zilber and Shelah can be seen as developing a formalism-free approach to model theory. Both Zilber's notions of a quasi-minimal excellent class and of a Zariski geometry, and Shelah's concept of an Abstract Elementary Class give axiomatic but mathematical definitions of classes of structures in a vocabulary τ. That is, the axioms are not properties expressed in some formal language based on τ but are mathematical properties of the class of τ-structures and some relations on it.

Of course, the *axiomatic method* has been entrenched in mathematical practice since the time of Euclid. Euclid's axioms were formulated in natural language, and the rules of proof were only implicit; in fact the codification of the rules was a long process which took place over centuries, culminating in the work of Frege and his successors.[4]

It is important to state at the outset that, unlike concepts like "truth" or "proof", formalism freeness is not a zero–one affair, generally, but a matter of

---

[4] Bourbaki distinguish between 'logical formalism' and the 'axiomatic method' e.g. in their oft-quoted remark, "We emphasize that it (logical formalism) is but one aspect of this (the axiomatic) method, indeed the least interesting one." See [27].

degree; a broad tendency of thought which is realised in different foundational and logical settings, sometimes very explicitly, and sometimes in a disguised form and taking place only in certain areas. Formalism freeness is not a sharp concept then, necessarily; as indeed the syntax/semantics distinction itself is not the sharpest of distinctions – or not always, as we noted.

There is an obvious sense in which formalism freeness, or on the other hand entanglement, turns on, or more precisely ramifies, the syntax/semantics distinction. In particular the question whether there can be implicit, ineliminable syntactic content in semantic frameworks, or conversely the question whether there is always an underlying semantic content in syntactic frameworks, is a central one for us. MacFarlane has written that

> The fact that certain logics can be formulated in completely syntactic terms, without reference to the meanings of their symbols, does not support the claim that they have no semantic content, as many have thought.

MacFarlane cites Cresswell as an example:

> One reason why "truths of logic" have been historically thought of in the narrow sense we have called "logical validity" is undoubtedly that the logically valid principles which result from treating only a small number of symbols as constants allow of reasonably simple formulation. It is even possible to formulate them without any reference to interpretation at all, a fact which is no doubt partly responsible for the idea that they are true independently of any content.[5]

Gödel defines the syntactic point of view in his draft "Is Mathematics a Syntax of Language?" as the view that mathematics is devoid of content. More precisely Gödel identifies the syntactic point of view with three assertions: First, mathematical intuition can be replaced by conventions about the use of symbols and their application. Second, "there do not exist any mathematical objects or facts", and therefore mathematical propositions are void of content. And third, the syntactical conception defined by these two assertions is compatible with strict empiricism.[6] Gödel argues for the ineliminability of the semantic content of a formalism, using the Second Incompleteness Theorem:

---

5   [44], p. 34.
6   [87], p. 320. See also "A philosophical argument about the content of mathematics," supplementary note to [127].

in order to establish the *consistency* of the system, one must import semantic intuition from outside the system.

Another formulation of what one might call the *semantic* point of view is due to Colin McLarty, who refers to the claim that the actual content of mathematics goes *beyond* any formalization as "expansive intuitionism", his term for Poincaré's reaction (or counterreaction) to formalism.[7] Yet another example of the semantic point of view, among the many that could be described here, is the abstract logic point of view due to Lindström and others, to be taken up in Section 2.2.2.

Of course the idea that mathematical reasoning could be in our sense "captured" by a formal system emerged very late. In particular, the notion of a signature emerged very late, as the question whether the inferential structure of mathematics could be replaced or modelled or expressed by a formalism – the locution here depending very much on one's philosophical perspective – became urgent around the turn of the twentieth century.

Setting aside the interesting mathematical questions – indeed the entirely new subject areas[8] that emerged from the various foundationalist programmes of the time – it is simply a fact that foundationalism in this form came and went with little lasting impact on mathematical practice. We touched on Burgess's notion of indifferentism in mathematics, indifferentism which takes various forms, e.g. indifference to the set-theoretical construction of the real numbers. As Burgess points out, there is also indifference to foundations überhaupt. The reasons for this are interesting, and though they are not strictly speaking the subject of this book, we can take note of certain responses at critical moments. For Kreisel, a sine qua non of interest in the Hilbert programme was not the consistency question per se but rather the presumed equivalence of second order consequence with, "at least in suitable contexts", formal derivability (as had been shown in the first order case).[9] However the question is of secondary importance for mathematics, concerned as mathematics is with (in Kreisel's terminology) *fundamental analysis*. "Logical hygiene", as he called it, might be useful, e.g. psychologically; but independence proofs, formalization, and such like, are not what mathematics deals with at its most fundamental level. "*C'est magnifique, mais ce ne sont pas les fondements*" he would say, of formalization.[10]

---

[7] See [176]. See also Detlefsen's [52].
[8] i.e. set theory, model theory, proof theory, etc.
[9] See [137], p. 146.
[10] Kreisel, [137].

2.1 Precedents

Our treatment of specific instances of formalism freeness begins in the next chapter with the development of computability, which was the paradigm example of formalism independence for Gödel in 1946. Subsequent chapters treat formalism freeness as it emerges in the work of Gödel and then of Tarski, also treating the various moves in this direction seen in contemporary foundational practice, driven by the work of those two figures.

Other historical developments are relevant from this point of view, but we will mention them only briefly here. We will consider the development of semantic methods in model theory in Chapter 6 — a development profoundly shaped by Tarski, whose semantic point of view was in turn rooted in the algebraic tradition in logic associated with Boole, Peirce and Schröder.[11]

Another precedent that is relevant from the suppression of the formalism point of view is represented by Brouwer, who prioritised mathematics over language altogether – in fact for Brouwer the radical decoupling of mathematics with language and its grounding in time intuition, was its fundamental feature. "Intuitionistic mathematics is an essentially languageless activity of the mind", he said, famously, in [28]. In particular, he expressed reservations about formalising intuitionistic logic. Brouwer's *First Ansicht* reads as follows:

> Completely separating mathematics from mathematical language and hence from the phenomena of language described by theoretical logic, recognising that intuitionistic mathematics is an essentially languageless activity of the mind having its origin in the perception of a move of time. This perception of a move of time may be described as the falling apart of a life moment into two distinct things, one of which gives way to the other, but is retained by memory. If the twoity thus born is divested of all quality, it passes into the empty form of the common substratum of all twoities. And it is this common substratum, this empty form, which is the basic intuition of mathematics.[12]

Fraenkel describes Brouwer's position in his [71] thus: "no formalized theory can do justice to intuitive (which is for them intuitionistic) mathematics or any of its subtheories". And Heyting, who axiomatised intuitionistic logic (against Brouwer's wishes), describes Brouwer's position this way:

---

[11] See for example https://plato.stanford.edu/entries/algebra-logic-tradition/
[12] [28].

... no formal system can be said to represent adequately an intuitionistic theory. There always remains a residue of ambiguity in the representation of the signs, and it can never be proved with mathematical rigour that the system of axioms really embraces every valid method of proof.[13]

Gödel's view of formalisation was complex. At various times it came close to Brouwer's in general outline, if not in the details. As he once said to Hao Wang: "the overestimation of language is deplorable".[14] On the other hand, Gödel would hardly have opposed the formalization of "the intuitive". As Gödel wrote to Wang in 1972, "Wittgenstein's negative attitude toward symbolic language is a step backward."[15]

As for the question of adequacy, or *faithfulness*, it was addressed by Gödel in different forms over his lifetime. To put it simply, whereas Brouwer saw what one might call the "faithfulness problem" as leading to a wholesale rejection of formal methods, for Gödel the faithfulness problem was rather the reason for the nonaxiomatisability of mathematics – an essentially critical view of formalisation as an *epistemic project*, though one which stops well short of Brouwer's position. We will return to the topic of Gödel and formalisation in Chapter 4.

Among logicians in whose work formalism freeness, in one form or another, played an explicit role, one should include Emil Post, who called for the return to meaning and truth in the opening of his [196], and the downplaying of what he called "postulational thinking". Post's concern to expose, unearth and otherwise make fully visible the line dividing "what can be done in mathematics by purely formal means [and] the more important part, which depends on understanding and meaning",[16] aligns him ideologically with Poincaré, in the latter's critique of logicism. As it turns out, Post's recommendation to develop recursion theory *mathematically*, by stripping off the formalism with which the theory was encumbered, led to the formalism free development of recursion theory just along the lines he advocated.[17] It also gave rise to the development of Post's own model of computability, namely Post Systems (see below).[18]

---

[13] [100], p. 102. Fraenkel comments on the above passage, that in spite of Heyting's "explicit disavowal ... A notion of (intuitionistic) truth can be satisfactorily defined for intuitionistic elementary logic under which the resulting formalized theory is complete and Heyting's logistic system is semantically complete." See [71], p. 323.

[14] See [280], remark 5.5.7.

[15] [280], p. 74.

[16] Gandy, [77], p. 93.

[17] As Kripke observed in his recent [144].

[18] In 1936 Post gave a similar analysis to that of Turing's, "with only Church [1936] to hand". See Kleene, [134].

2.2 Entanglement and Formalism Freeness: Varieties

Before closing this chapter we offer a few concrete examples of entanglement and formalism independence/formalism freeness, among the many that we could have chosen. Other examples are treated separately in subsequent chapters.

Our notion of entanglement is that it occurs whenever changes in a given formalism or logic, on the level of syntax, of quantification, or even involving a choice of semantics, yield an impact in the formal environment. Thus a change of logic, a change of formalism or a change in the syntax can create entanglement. Examples cited above include the 0–1 laws for finite structures, and the Bernays–Schoenfinkel–Ramsey class. We also considered the reals in the context of adding the sine function, or in turn a predicate for the natural numbers, to the axioms for the complete ordered field.

2.2.1 Examples from Set Theory

The constructible hierarchy, denoted L,[19] provides an instance of the entanglement of a canonical structure with a *logic*; from another point of view L proves to be to some degree formalism free. Regarding entanglement: In its original presentation L was built using first order logic. As Myhill and Scott [187] showed, if L is constructed using second order logic, the class obtained is HOD, the hereditarily ordinal definable sets.[20] In this sense L is sensitive to the underlying logic. Other examples of this sensitivity to the underlying logic are given in Section 4.4.

To what extent is this sensitivity a fundamental feature of L? As we will point out in Chapter 4, while L is indeed sensitive to perturbations in the underlying logic, at the same time L is extremely robust, in that there is a very large class of logics that L "reads" as first order. That is, building L with a logic from this class yields the same inner model as one would obtain if one used first order logic. This fact is interesting in that this class of logics which L reads as first order includes those such as Magidor–Malitz logic, assuming the set-theoretical assumption $0^{\#}$. By the Lindström characterisation this logic is far from being first order as it is (consistently) badly incompact. On the

---

[19] See [116] p. 28 for the definition of the constructible hierarchy.

[20] A set a is *hereditarily ordinal definable* if a itself and also every element of the transitive closure of a is ordinal definable. Both $V = \mathrm{HOD}$ and $V \neq \mathrm{HOD}$ are consistent, relative to the consistency of ZFC. The first result is due to Gödel [80] and the second is due to Cohen [41].

other hand logics which are close to being first order according to the Lindström characterisation yield L a new inner model. Thus L, in its "readings" of logics, does not track the Lindström characterisation of first order logic. (See Section 4.4.)

HOD is similarly robust, as we will show. Note that L, unlike HOD, is also absolute in a very strong sense.[21] In Chapter 4 we will discuss extensions of the Myhill–Scott result to other logics.

Looking at set theory on a more basic level, the working set theorist is often indifferent to the syntactic form of canonical objects of study – though just as often the syntactic form of concepts is very important in set-theoretic practice, witnessing the entanglement of the concept with the relevant syntax. For example of the former, consider the concept of ordered pair, which is often cited as manifesting (in set theory) an artificial dependence on notation, namely Kuratowski's notation $\langle a, b \rangle = \{\{a\}, \{a, b\}\}$. That the accusation is based on a misunderstanding – the set theorist's dependence on the notation is not a genuine one – does not make the example less illuminating. What is at issue is the idea that concepts in set theory, such as the ordered pair concept, are defined uniquely and not up to isomorphism.[22] However this does not indicate an attachment to any particular way of presenting the unique object in question. Set-theoretic practice is invariant under the choice of how things like ordered pairs are actually defined. In the end, the set theorist's reliance on the formalism $\langle a, b \rangle = \{\{a\}, \{a, b\}\}$ extends only as far as the recognition that the ordered pair can be so defined. But then the formalism is ignored, having served its purpose.[23]

There is a point to be made here about definability in the case of the ordered pair and similar concepts, which is that the set-theoretic definition of such basic concepts is usually absolute for transitive models of set theory,[24] being of a Δ_0

---

[21] Since L is a model of ZFC, the construction of L can be carried out inside L. But one obtains by doing this exactly the same structure. This is not true of HOD. For example, if $0^\sharp$ exists, it is an element of HOD but not of L, thus the HOD of L is different from the actual HOD. In fact, HOD is non-absolute in the even stronger sense that HOD may not satisfy $V = \mathrm{HOD}$. See [174].

[22] This is in contrast to, for example, definability in second order logic. Speaking very roughly, in second order logic definability is done by means of concepts that are specified up to isomorphism only. Of course one can define notions up to isomorphism in set theory, but one is not forced to, as in second order logic.

[23] The exact form of the ordered pair comes up only in some rather trivial rank-computations in set theory.

[24] Given a theory T a statement in a vocabulary including $\in$ is absolute if and only if it is provably Δ_1 in the language of T; that is, provably equivalent to both a Σ_1 and a Π_1 formula in the given language. If a sentence is absolute, then if it is true in a model M of the theory it is true in all end extensions of M, being equivalent to an existential formula; and

form, i.e. involving only bounded quantifiers. It is important for set-theoretic practice that concepts of this kind are absolute.

G. Sacks [212] sees the absoluteness of the central notions of *model theory* as the reason why model theory need not concern itself with set-theoretic assumptions involving e.g. cardinality – a somewhat different form of indifferentism:

> B. Dreben ... once asked ... "Does model theory have anything to do with logic?" It is true that model theory bears a disheartening resemblance to set theory, a fascinating branch of mathematics with little to say about fundamental logical questions, and in particular to the arithmetic of cardinals and ordinals. But the resemblance is more of manners than of ideas, because the central notions of model theory are absolute, and absoluteness, unlike cardinality, is a logical concept.[25]

We will return to the possible set-theoretic content of model theory in Chapter 5.

A second example of formalism freeness from set theory has to do once again with Gödel's constructible hierarchy. Initially the hierarchy was given in terms of explicit first order definability [80]:

$$
\begin{aligned}
L_0 &= \emptyset \\
L_{\alpha+1} &= \mathrm{Def}(L_\alpha) \\
L_\nu &= \bigcup_{\alpha<\nu} L_\alpha,
\end{aligned}
$$

where $\mathrm{Def}(L_\alpha)$ consists of all the sets that are first order definable with parameters in L_α. In 1940 Gödel gave a second presentation of the constructible sets, as the closure of the class of ordinals under the so-called "Gödel operations" [81]. The definition of the L-hierarchy is obtained by closing the ordinals under the functions:

$$
\begin{aligned}
\mathcal{F}_1(X,Y) &= \{X,Y\}, \\
\mathcal{F}_2(X,Y) &= \{(a,b) \in X : a \in b\}, \\
\mathcal{F}_3(X,Y) &= X - Y, \\
\mathcal{F}_4(X,Y) &= \{(a,b) \in X : b \in Y\}, \\
\mathcal{F}_5(X,Y) &= \{b \in X : \exists a (a,b) \in Y\}, \\
\mathcal{F}_6(X,Y) &= \{(a,b) \in X : (b,a) \in Y\}, \\
\mathcal{F}_7(X,Y) &= \{(a,b,c) \in X : (a,c,b) \in Y\}, \\
\mathcal{F}_8(X,Y) &= \{(a,b,c) \in X : (c,a,b) \in Y\}.
\end{aligned}
$$

This latter, in Gödel's words, more perspicuous presentation, is given in natural language, with no satisfaction or definability predicates occurring in it. In

conversely if it is true in a model M then it is true in all submodels of M, as it "persists downwards", being equivalent to a universal formula. See [59].

[25] Sacks, ibid, p. 199.

particular, the Gödel functions belong to natural language, as do the ordinals, if one wishes to think of them as given informally. In his Brown Lecture [*1940a] Gödel says that this proof is "more perspicuous", also that it is more in line with "Hilbert's Program".[26] As to our particular concern here Gödel seems motivated by considerations which are strikingly close to those which motivated Tarski in his programme to replace the concept of first order definability by a so-called "mathematical" analogue, or more broadly to avoid metamathematical concepts altogether, in order to appeal to the working mathematician. We will return to this point in Chapter 5.

2.2.2 Abstract Logics; Semantic Characterisations of Metamathematical Concepts

Defined in very general terms, a logic or an *abstract logic* is a triple $L = (S, F, \models_L)$ where $\models_L \subseteq S \times F$. Elements of the class S are called the structures of L, elements of F are called the sentences of L, and the relation $\models_L$ is called the satisfaction relation of L. Usually S is the class of ordinary structures i.e. non-empty sets with relations, functions and constants on them. Then it is almost a rule to demand that if $M \cong N$, then for all $\phi \in F$:

$$M \models \phi \iff N \models \phi.$$

P. Lindström's original characterisation [152] of an abstract logic associates formulas directly with model classes,[27] where satisfaction is replaced by membership in the relevant class. An abstract logic in Lindström's original sense is then in effect a family of model classes closed under some simple operations, such as permutation of symbols, conjunction, existential quantifier, etc.[28] The notion of syntax is not part of the definition of an abstract logic, so strictly speaking these are not formalisms in the sense given above.

In detail: we an *abstract logic L** to be a family[29] of model classes satisfying the following simple closure properties:

- If $K \in L^*$, $K' \in L^*$ and K, K' have the same similarity type then $K \cap K' \in L^*$.

---

[26] In his introduction to [*1940a] Solovay rejected Gödel's reference to the Hilbert programme in this context.
[27] Recall that a model class is a class of models of the same similarity type, closed under isomorphism.
[28] Lindström later gave a syntactic definition, see [154].
[29] A family $\{K_i : i \in I\}$ of classes, indexed by the class I, can be identified with the class $\{(i, x) : i \in I \land x \in K_i\}$.

- L^* is *closed under negation*, i.e. if for all $K \in L^*$ there is $K' \in L^*$ such that a model of the appropriate similarity type is in K' if and only if it is not in K.
- L^* is *closed under existential quantification*, i.e. if $K \in S$ and c is a constant symbol in the vocabulary of K, then there is $K' \in L^*$ (without the constant symbol c) such that a model is in K' if and only if it can be expanded to a model in K by adding value to c.
- L^* is *closed under renaming*, i.e. whenever π is a permutation of relation and constant symbols (we assume we do not have function symbols, for simplicity) which respects arity, and we extend π in a canonical way to structures, then for all $K \in L^*$, there is $K' \in L^*$ such that a model M is in K' if and only if its image under π is in K.
- L^* is *closed under free expansions*, i.e. whenever $K \in L^*$, and we have a similarity type τ bigger than the similarity type of K, there is $K' \in L^*$ of similarity type τ such that K is the class of reducts of models in K'.[30]
- The following is a notion of equivalence: An abstract logic L^* is a *sublogic* of another abstract logic L^{**}, in symbols $L^* \leq L^{**}$, if every $K \in L^*$ satisfies $K \in L^{**}$. If $L \leq L'$ and $L' \leq L$, we say that L and L' are *equivalent*, $L \equiv L'$.

The set of all first order definable model classes is an abstract logic in this sense, as the above closure properties are clearly satisfied. Other abstract logics arise from infinitary languages, generalized quantifiers, higher order logic and combinations of such.

The terminology, once again: if a model class $K \in L^*$, then K is called a "sentence" of L^*, and if $M \in K$, then M is called a "model" of K. In infinitary logic it may be thought to be more natural to think of sentences as infinite trees, a departure from the idea of a sentence as a string of symbols subject to recursive formation rules.[31] In abstract logic according to the above definition the departure is complete.

---

[30] If M is a model with vocabulary L and $L_1 \subseteq L$, then the *reduct* $M \restriction L_1$ of M to the vocabulary L_1 is the model M_1 with the same domain as M and the following structure: $R^{M_1} = R^M$ if $R \in L_1$, $f^{M_1} = f^M$ if $f \in L_1$, and $c^{M_1} = c^M$ if $c \in L_1$.

[31] As Keisler writes:

All this is done within set theory. For definiteness, we agree that each variable v_α is the ordered pair $\langle v, \alpha \rangle$, and each of the other symbols of L is a natural number. Atomic formulas are finite sequences of symbols. ... Thus each formula of $L_{\omega_1, \omega}$ is a set.

Viewing this family of sets under the subformula relation, the set of formulas, construed as finite sequences of sets, forms naturally a tree. See [122], p. 6.

This very general definition of a logic admits a natural formulation of such basic properties of logics as the Compactness Theorem: If a set[32] of "sentences" (i.e. model classes) of L^* is such that every finite subset has a model, then the whole set has a model. The Downward Löwenheim–Skolem Theorem is formulated as follows: If a sentence (i.e. a model class) in L^* has an infinite model (in it), then it has a countable model (in it).

Lindström famously proved that first order logic is characterised by these two properties in his [153]. Similarly, Lindström's Theorem characterising first order logic as the maximal logic which satisfies two model-theoretic properties, the Downward Löwenheim–Skolem Theorem and the Compactness Theorem, means that one can view first order logic purely semantically with no concern as to the syntax. As long as these two model-theoretic properties are satisfied, the concept of a definable model class is the same. As far as definability of model classes is concerned, first order characterisability manifests then, thanks to Lindström's Theorem, very strong formalism freeness.[33]

Barwise finds Lindström's theorem "unsatisfactory" because it suppresses what appears to be an implicit syntax:

> Lindström avoids syntactic considerations altogether since he deals directly with classes of structures, rather than with the sentences which define them. We find this approach unsatisfactory on two grounds. In the first place, it seems contrary to the very spirit of model theory where the primary object of study is the relationship between syntactic objects and the structures they define. Secondly, it fails to make explicit that the closure conditions on the classes of structures (like formation of indexed unions and its inverse) arise out of natural syntactic considerations, considerations which seem implicit in the very idea of a model-theoretic vocabulary.[34]

We will return to the notion of implicit syntax in Chapter 5.

In fact, many other logics permit a semantic characterisation. Barwise[35] showed that if $\kappa = \beth_\kappa$, then $\mathcal{L}_{\kappa\omega}$ is the maximal logic which has the Karp Property[36] and well-ordering number $\leq \kappa$.[37] This is not as sharp a

---

[32] A *set* $\{K_i : i \in I\}$ of model classes, indexed by the *set* I, can be identified with the class $\{\langle i, x \rangle : i \in I \land x \in K_i\}$.

[33] One might also mention in this connection the result of J. Akkanen, that a model class in a finite vocabulary is definable in first order logic if and only if it is definable in KPU^- (Kripke-Platek set theory with urelements and without the infinity axiom). This gives an alternative set-theoretic characterisation of first order logic. See Akkanen, [3].

[34] [17].

[35] See [17].

[36] Partially isomorphic structures are elementarily equivalent with respect to the logic.

[37] This is the smallest ordinal α such that if a sentence of the logic has only models in which a binary predicate R is well-ordered, then in every model the order type of R is $< \alpha$. See [17].

semantic characterisation as Lindström's, although Barwise relies heavily on Lindström's proof; but the two conditions are in principle comparable to the conditions occurring in Lindström's proof, namely the Löwenheim–Skolem and Compactness Theorems. Thus $\mathcal{L}_{\kappa\omega}$, for $\kappa = \beth_\kappa$, manifests some extent of formalism freeness, in the sense that its class of definable model classes is not entirely entangled with the ordinary syntax of $\mathcal{L}_{\kappa\omega}$. Another interesting case is the recent logic L^1_κ of Shelah [228] (see below), which is between $\mathcal{L}_{\kappa\omega}$ and $\mathcal{L}_{\kappa\kappa}$ and satisfies a Lindström Theorem. Yet another class of infinitary languages which demonstrates remarkable formalism freeness, also due to Barwise [16], is the family of admissible fragments $\mathcal{L}_A$ of $\mathcal{L}_{\infty\omega}$. If A is an admissible set, then the language $\mathcal{L}_A$ is maximal among logics which are the intersection of A with a strictly absolute logic.[38] So the syntax of $\mathcal{L}_A$ does not "matter" as long as it is strictly absolute.[39]

Returning to first order logic, it is interesting to note that the principal meta-mathematical theorems concerning this logic permit a semantic reformulation. The syntactic version of the Craig Interpolation Theorem is:

Theorem 2.2.1 (Craig) *Suppose ϕ_1 is a first order sentence with vocabulary L_1 and ϕ_2 is a first order sentence with vocabulary L_2. If $\models \phi_1 \rightarrow \phi_2$, then there is a first order sentence (the "interpolant") ψ with vocabulary $L_1 \cap L_2$ such that $\models \phi_1 \rightarrow \psi$ and $\models \psi \rightarrow \phi_2$.*

Here is a purely semantic version of the theorem:

Theorem 2.2.2 (Craig) *Let K_1 be a model class with vocabulary L_1, and K_2 be a model class with vocabulary L_2. Suppose $K_1^* = K_1 \restriction L_1 \cap L_2$ is the class of reducts $M \restriction L_1 \cap L_2$ of models $M \in K_1$. Similarly $K_2^* = K_2 \restriction L_1 \cap L_2$. Then if K_1 and K_2 are first order definable model classes and $K_1^* \cap K_2^* = \emptyset$, then there is a first order definable model class K in the vocabulary $L_1 \cap L_2$ such that $K_1^* \subseteq K$ and $K_2^* \cap K = \emptyset$.*

Note that K_1^* and K_2^* need not be first order definable. They are *projections* of first order definable model classes K_1 and K_2. Tarski calls them PC classes (PC for projective class), while first order definable model classes are referred to as EC classes (E for elementary). So interpolation says that any two disjoint PC classes can be separated by an EC class.

---

[38] I.e. sentencehood is Σ_1^{KP} and the satisfaction predicate is Δ_1^{KP}.
[39] We are grateful to Jouko Väänänen for suggesting the examples in this paragraph.

2.2.3 Logics Given by a Game, Logics without Syntax

One way to view a *logic* in a formalism free manner is through the concept of a game. The use of games in mathematics has a provenance dating back to the Lwów School of the 1930s and 40s (if not earlier). Problem 43 in the so-called Scottish Book involved the Banach–Mazur game; in 1962 Mycielski and Steinhaus introduced the Axiom of Determinacy (AD), an important set-theoretic principle based on two-person games in Baire space.

Some terminology: a *strategy* for a player is a function that determines exactly how that player should move, depending on the history of the game to that point, i.e. depending on the earlier moves of the two players. It is a *winning strategy* if the player wins no matter what moves the other player makes. Now if A is a set of real numbers, or equivalently a subset of ω^ω, the game associated to A is as follows: two players I and II take turns choosing natural numbers:

I	x_0		x_1		$\cdots$	x_n	$\cdots$
II		y_0		y_1	$\cdots$		$y_n\cdots$

Player I wins the game if the infinite sequence $\langle x_0, y_0, x_1, y_1 \ldots\rangle$ belongs to A.

We say that the set A is *determined* if this game is determined, meaning that one of the players has a winning strategy. The Axiom of Determinacy AD states that every subset of ω^ω is determined. AD is incompatible with the Axiom of Choice but holds in $L(\mathbb{R})$, assuming large cardinals. Closed sets are determined, and Borel sets are determined, a result due to D.A. Martin [171].

The classical descriptive set theory project of the 1920s and 30s, which sought to classify point classes of the descriptive hierarchy, so-called, came to a halt in the 1930s, with Luzin famously predicting that "we will never know the answer to the measure question for the projective sets" [155]. In an astonishing achievement due to Martin and Steel [173] the problem was finally solved: projective sets are determined, assuming large cardinals.[40]

It turns out that the concept of elementary equivalence for *logics* can be presented via the concept of a game: Fraïssé gave conditions characterising first order elementary equivalence;[41] Ehrenfeucht rephrased Fraïssé's conditions in terms a game. These came to be known as Ehrenfeucht–Fraïssé or EF-games, and they involve two players playing on two first order structures M and M' (without function symbols, for simplicity), with each player choosing elements from M and M' in turn. More precisely, in the game $EF_n(M, M')$ of length n,

---

[40] The projective sets are obtained from closed sets by iterating taking complements and continuous images.

[41] [73].

given a position in the game, if player I chooses an element a_i from M (M'), the player II must choose an element b_i from M' (M). The intuition behind the game is that player II attempts to prove that the structures are very similar, if not isomorphic; player I wishes to show the contrary. Player II wins the game if the partial mapping $a_i \leftrightarrow b_i$ constructed during the game is a partial isomorphism, otherwise player I wins. The relevant theorem is:

Theorem 2.2.3 *Let $M \sim_n M'$ mean the second player has a winning strategy in the EF-game of length n on M and M'. Let $M \equiv_n M'$ mean that exactly the same first order sentences of quantifier rank $\leq n$ are true in the models M and M'. Then $M \equiv_n M'$ iff $M \sim_n M'$. Moreover, a model class in a finite relational vocabulary is first order definable if and only if there is an n such that it is closed under $\sim_n$.*

Other logics for which the notion of elementary equivalence can be characterised in terms of games rather than formulas include the following:

1. Suppose Q is a monotone generalized quantifier in the sense of [152]. The EF-game can be modified from first order logic to the extension $L(Q)$ of first order logic by Q, and then the above results still hold, i.e. if $A \sim_n^Q B$ is the resulting equivalence relation, then:
   Fact: A model class K is $L(Q)$-definable if and only if there is n such that K is closed under $\sim_n^Q$. [55]. In consequence, $A \equiv_{L(Q)}^n B$ iff $A \sim_n^Q B$. [281].

2. Let $A \sim_\alpha B$ if the second player has a winning strategy in the EF-game of length ω on A and B, when α is the clock (the first player goes down the ordinal α, so the game is always finite).
   Fact: A model class K is definable in $L_{\infty\omega}$ iff there is α such that K is closed under $\sim_\alpha$. Two models A, B satisfy the same sentences of $L_{\infty\omega}$ of quantifier rank less than or equal to α if and only if $A \sim_\alpha B$. [117].

3. Let $A \sim_\theta^0 B$ if the second player has a winning strategy in Shelah's game as in [228], where players play sets of size $\leq \theta$ according to certain rules. The relation $\sim_\theta^0$ is not known to be transitive. Let $\sim_\theta^1$ be the closure of $\equiv_\theta^0$ to a transitive relation. Then $\sim_\theta^1$ is an equivalence relation.
   Shelah's [228] logic L_κ^1, $\kappa = \beth_\kappa$, where κ is of cofinality ω, is defined by declaring that the definable model classes are the model classes that are closed under $\equiv_\theta^1$ for some $\theta < \kappa$, i.e. elementary equivalence in Shelah's logic L_κ^1 is $\sim_\kappa^1$. Shelah proves a Lindström type characterisation of this logic. Interestingly, Shelah does not actually give a syntax for this logic, and perhaps the syntax is not "needed" in order to characterise the logic completely, because there is already a semantic characterisation, via its

Lindström theorem. L_κ^1 is also a very interesting "logic" because it satisfies interpolation, and yet, again, it has no syntax at all.

Presenting a logic by a game is to give a semantic characterisation of the concept of elementary equivalence relative to the logic. Concretely, the characterisation of elementary equivalence for a logic induces an equivalence relation on the space of models in the following way: two structures of the same similarity type are equivalent if they satisfy the same sentences of the logic. Does giving a criterion for elementary equivalence explain/capture the character of a logic? If, as it has often been said, logic is a tool,[42] here one might say the logic is a tool for distinguishing among structures. If in the most general terms a logic is given by its syntax, i.e. a notion of a formula of the logic, and its satisfaction relation, then a characterisation of elementary equivalence for the logic in this game sense apparently gives us no information about its syntax. But we saw that some logics *have no syntax at all*, e.g. Shelah's logic L_κ^1, about which there is the question whether the syntax is really needed in order to understand the character of the logic, as the Lindström Theorem for the logic defines it uniquely. Other logics have a syntax in only a weak sense, i.e. depending on the background set theory (see [260]). From another point of view it an important property of the logic, that it has a syntax.

We said that the characterisation of elementary equivalence for the logic given by the relevant game gives us no information about the syntax of the logic, apparently. But in fact one *can* always recover a logic from a game, albeit in a rather artificial way. How is this done? One considers the class of structures for the game, structures in a given vocabulary. One then defines an equivalence relation on the class, so that two structures M and M' from the class are equivalent if player II has a winning strategy in the game played on M and M'. One can then associate a sentence ϕ_M to each equivalence class $[M]$ and a notion of satisfaction $\models_M$ so that $M' \models \phi_M$ if and only if $M' \in [M]$. See M. Nadel's [188] for details.

What *is* a logic, after all? Nadel expresses the difficulty of the question:

> Implicit in the above questions lies the more basic question of deciding upon exactly what constitutes a logic. In the case of a particular example such as Shelah's,[43] this issue can be sidestepped, as the particular example may be generally recognizable as a logic. In particular, any formal definition of logic would have to include it. However, in our more general setting, the issue seems difficult, if not impossible, to avoid.

---

[42] See e.g. thesis 1 of [10], p. 3.

[43] Nadel is referring here to a different logic than L_κ^1.

Though this question certainly has a direct bearing on the present paper, no attempt will be made herein to fully resolve it. Instead we will work in the context of pre-existing definitions. It is hoped that any dissatisfaction with the examples presented below may give some insight into possible changes or improvements in these current notions.[44]

In Section 2.2.2 we defined the concept of an *abstract logic* as a triple $L = (S, F, \models_L)$, where $\models_L \subseteq S \times F$, where S are the *structures* of L, F are the *sentences* of L, and the relation $\models_L$ is the *satisfaction relation* of L. The abstract logic point of view may appear to leave the philosophical question "what is a logic?" or perhaps more fundamentally, the question "what is inference?" untouched. In an abstract logic the attempt is to capture logical concepts semantically, i.e. by studying model classes directly, and up to isomorphism. The emphasis here is on explaining the truth of the sentence, where the "meaning" of a sentence is taken to be its model class. This is opposed to e.g. a concept of meaning that behaves compositionally, e.g. in which the meaning of a proposition is derived from its atomic components.

2.2.4 A Simple Preference for Semantic Methods?

In his logical work Tarski pursued an explicit programme to replace meta-mathematical notions with what he called "mathematical" ones. For example he proved that a class of structures in a finite relational language is universally axiomatisable if and only if it is closed under isomorphism, substructure and if every finite substructure of a structure $A \in K$, then $A \in K$.[45]

In his [9] and [10] Baldwin notes the distinction between a *mathematical* as opposed to *logical* property:

> An inquiry can be 'formalism-free' while being very careful about the vocabulary but eschewing a choice of logic and in particular any notion of formal proof. Thus it studies mathematical properties in [Tarski's JK] sense ... [46]

Much of the model theory in the 1950s involved the search for *mathematical* descriptions of canonical model classes. We will return to these particular model-theoretic issues in Chapter 5, in which we explore Tarski's programme at length.

---

[44] ibid, p. 103.

[45] See [250]. A class of structures is universally axiomatisable if it is definable in first order logic by sentences in prenex form with the quantifier prefix containing only universal quantifiers.

[46] [9], p. 97.

2.3 Natural Language Moves

In the coming chapters we will consider various moves made in the direction of suppression of logic and syntax in foundational and model-theoretic practice – natural language moves, as they could be called, made by the early workers on computability, by Gödel, by Tarski and by contemporary logicians such as Shelah and Zilber. We set the table with a few key examples, indicating thereby what it means to survey logical space through this particular lens.

As we see, formalism freeness shows up in a number of different ways. We considered transcendence or robustness with respect to a logic, as evidenced by the fact that for example the constructible hierarchy can be built using many different logics (in addition to the, usually, original first order one). There is also stability across a class of formalisms, viz computability. We also considered admitting a *mathematical* (in Tarski's terminology) rather than a logical treatment, as exemplified by constructibility given in terms of Gödel functions. Tarski's characterisation of universal axiomatisability given above[47] is another example of this kind of formalism freeness, as is any reformulation in semantic terms of syntactic and/or metamathematical concepts. We saw indifference to foundational framework decisions, in this case indifference with respect to a choice of semantics (Henkin or full) in the second order context. Formalism freeness can also refer to the absence of a formal proof system, as in Baldwin's [10]:

> But he [i.e. Hilbert JK] has also given a 'formalism-free' proof of Theorem 12.4.2.2. ... In the weakest sense this proof is formalism-free since there is no formal proof system. But even more, the 'interpretation' [in question JK] makes no reference to the formal vocabulary or any notion of a formal language. But as we noted in proving Theorem 12.4.4, this formalism-free proof translates to the existence of a formal proof by the extended completeness theorem. (Of course, this translation was not available to Hilbert in 1900.) That is, in the proof of the formal Theorem 12.4.2.2, we invoked Hilbert's semantic proof of Theorem 12.4.2.1. Such a translation is a standard consequence of the extended completeness theorem and a routine model-theoretic tool. This does not mean that the semantical proof is tacitly formal; this translation just expresses the content of the completeness theorem.[48]

One can think of these various modes of formalism freeness as the simple preference for semantic methods, e.g. methods which do not involve or require the specification of a syntax or even a logic – at least not prima facie. The

---

[47] i.e. in terms of closure under isomorphism, substructure and the property that if for every finite substructure B of a structure A, $B \in K$ then $A \in K$.

[48] ibid, pp. 269–270.

fact that semantic concepts can be construed syntactically does not hinder our investigations here, it rather supports them; nor does the question, what is our knowledge of semantics based on, if not on some formalism? In fact the answer to this question is the heart of our approach. We take it as given that mathematics has – and has always had – semantic content, and the question is, whether and to what extent the various relevant formalisms capture it.

For logicians, formalism freeness may appear somewhat elusive, given the apparent entanglement of natural language mathematical discourse with the various formalisms that have emerged to date – formalisms which leap into view for logicians; formalisms which succeed, after all, in codifying almost the entire inferential structure of mathematics.

Codifiability is important. Our interest here is in the "side effects" of codi-fication, in what is left out; in what has to be said or done in natural language. Both Gödel and Tarski reverted to natural language in pursuit of certain programmatic goals. For Gödel, who sought decidability in mathematics, cod-ifiability was irrelevant. For Tarski, as we will see in the coming chapters, the pursuit of codifiability, or more broadly of metamathematics generally, amounted to a form of self-exile, exile from the domain of mathematics proper. In both cases one sought a natural, unentangled *mathematical* framework, even in the midst of foundational practice.

We turn in the next chapter to the paradigm example of what Gödel called formalism independence, namely the concept of a Turing Machine.

3

Computability: The Primary Example

I hold up my hand and I count five fingers. I take it on faith that the mapping from fingers onto numbers is *recursive* in the sense of the mathematician's definition of the informal concept, "human calculability following a fixed routine". I cannot *prove* the mapping is recursive – there is nothing to prove! Of course, mathematicians can prove many theorems about recursiveness, moving forward, so to speak, once the definition of the concept "recursive" has been isolated. Moving backwards is more difficult and this is as it should be: for how can one possibly hope to prove that a mathematical definition captures an informal concept, in this case human calculability?

We introduced the term "formalism freeness" in the previous chapter to refer to a complex of ideas involving the suppression of syntax and the forefronting of semantics in foundational practice. In this chapter we will introduce the first of our three "case studies", episodes in foundational practice in which formalism freeness plays a central role. We have said that for Gödel, computability was the paradigm example of a formalism independent concept. Here we will track the emergence of the concept of computability in the 1930s, from Gödel's perspective, mainly, and from the formalism freeness point of view generally.

Consider the concept of *adequacy*, or more precisely the relation "x captures y." The question Gödel posed in his 1946 Princeton Lecture was the following: if y is taken to be computability, definability or provability, does there exist an adequate and unique choice of x?[1] We will argue in this chapter that for Gödel, adequacy and formalism freeness are tightly related (at least in the case of computability) in the sense that an adequate concept of "computable function" is necessarily formalism free.

---

[1] See also Post's Collected Works [197]. See also A. Urquhart's [258], p. 469.

3.1 On Adequacy

By which criteria do we shape, commit ourselves to, or otherwise assess standards of adequacy? This is the problem of *faithfulness*; the problem of what is lost whenever an intuitively given mathematical concept is made exact, or, beyond that, formalised; the problem, in words, of the *adequacy* of our mathematical definitions. It is always present in mathematics, but it is a philosophical problem rather than a mathematical one, in our view, as there is nothing in this idea of "fit" that can be subject to mathematical proof.[2] Of course the faithfulness problem is only a problem for those who take the view that a sharply individuated concept is present, intersubjectively, in the first place. In his [107] Hodges takes a different view: "*Turing's Thesis* is the claim that independent of Turing we have an intuitive notion of an effectively computable function, and Turing's analysis exactly captures this class of functions ... This kind of claim is impossible to verify. Work like Turing's has a power of creating intuitions. As soon as we read it, we lose our previous innocence."[3]

Over time, logicians have developed a number of what one might call coping strategies. For example, some have raised the question of intensional adequacy in connection with the Second Incompleteness Theorem, as establishing the theorem may turn on the *meaning* of the (formal) consistency statement as read by the relevant theory.[4] On this view, briefly, one should grant the meta-theoretical claim that a theory T cannot prove its own consistency only when there is a sentence which T both "recognises" as a consistency statement, and which T cannot prove. In the literature a theory T is thought to "recognise" its consistency statement, just in case it can be proved in T that its proof predicate satisfies the three Hilbert–Bernays derivability conditions.[5] A consistency statement is generally a statement to the effect that a sentence of some kind – or of any kind – is not provable. Because of this the adequacy claim in question will then rest on the further claim that the provability predicate of T is recognised by T as adequately representing "genuine" provability.

Another coping strategy involves the use of the word *thesis* – a word that serves to flag an adequacy claim, if not to bracket it. The Church–Turing Thesis, for example, in its present formulation, equates the class of intuitively

---

[2] Some would claim to the contrary that Turing proved a theorem equating human effective calculability with Turing calculability. See below.

[3] ibid, p. 489.

[4] See Feferman [58], Detlefsen [53], Franks [74] and Pudlak [198]. The situation with the Second Incompleteness Theorem is in contrast to the First Incompleteness Theorem, which has nothing to do with the *meaning* of the undecidable sentence $\mathcal{G}$. All that matters in that case is exhibiting a sentence $\mathcal{G}$ that is undecidable, regardless of what it "says".

[5] As given in their 1934 [101].

computable number-theoretic functions with the class of functions computable by a Turing Machine.[6] Other conceivable theses in mathematics include *Weierstrass's thesis*, as some have called it,[7] namely the assertion that the $\epsilon - \delta$ definition of continuity correctly and uniquely expresses the informal concept; *Dedekind's thesis*, asserting that Dedekind gave the correct definition of the concept "line without gaps";[8] or, alternatively, asserting that Dedekind gave the correct definition of "natural number"; the *area thesis*, asserting that the Riemann integral correctly captures the idea of the area bounded by a curve. *Hilbert's Thesis*, a central though contested claim in foundations of mathematics, refers to the claim that "the steps of any mathematical argument can be given in a first order language (with identity)".[9]

To many working mathematicians the correctness of theses such as Weierstrass's is, in Grothendieck's sense of the word, obvious – what more is there to say about continuity than what is encapsulated in Weierstrass's definition of it? The possibility that meaningful concepts may arise in future practice, which may fall under the intuitive notion, while not fitting, e.g. Weierstrass's definition, is simply not considered. In such a case one speaks of theorems, not theses. Turing, for example, is thought by some not to have formulated a thesis but rather to have proved a *theorem* relating human effective computability to computability by means of a Turing Machine. *Turing's Theorem* is defined by Gandy, for example, as follows: Any function which is effectively calculable by an abstract human being following a fixed routine is effectively calculable by a Turing Machine – or equivalently, effectively calculable in the sense defined by Church – and conversely.[10]

In sum, there are "theses everywhere", as Shapiro notes in his [219]; not necessarily provable, but at the same time in most cases no longer subject to doubt.

How does the mathematician cope with the problem of adequacy? We quoted Wimsatt: "Things are robust if they are accessible (detectable, measurable, derivable, definable, producable, or the like) in a variety of

---

[6] This is "Thesis T" in Gandy [1988]. A number of writers on computability take the Church–Turing Thesis to have been proved. See Soare's [241] and Gandy's [77]. The intuitively computable number-theoretic functions are henceforth referred to as "effectively computable."

[7] See Shapiro, [219].

[8] See [233].

[9] Kripke, [144], p. 81. See also Burgess's [30], Rav's [205] and Shapiro, [219]. The terminology "Hilbert's Thesis" may be slightly misleading, given Hilbert's use of a second order axiomatisation in *Die Grundlagen der Geometrie* [103]. Thanks to J. Baldwin for pointing this out.

[10] [77], p. 83.

independent ways."[11] But there is also *grounding*: the idea that, among the class of conceptually distinct precisifications of the given (intuitive) concept, one of them stands out as being indubitably adequate – as being the *right* idea. This happened, evidently, with the notion of "finite". There are a number of (extensionally) equivalent definitions,[12] but the notion that a set is finite if it can be put into one to one correspondence with a natural number seems to be what "every" mathematician means by the term "finite" – even though the definition is blatantly circular, on its face.[13] This also happened after 1936 in connection with the notion of computability, when the Turing analysis of human effective computability was taken by logicians to have solved the adequacy problem in that case.

The number theorist Michael Harris uses the word "avatar" to express a condition of *no grounding*, or *in*adequacy – and within that, the condition of knowing *that*, but not *why*:

> I suggested that the goal of mathematics is to convert rigorous proofs to heuristics – not to solve a problem, in other words, but rather to reformulate it in a way that makes the solution obvious ... "Obvious" is the property Wittgenstein called *übersichtlich*, surveyable. This is where the avatars come in. In the situations I have in mind, one may well have a rigorous proof, but the obviousness is based on an understanding that only fits a pattern one cannot yet explain or even define rigorously. The available concepts are interpreted as the avatars of the inaccessible concepts we are striving to grasp.[14]

In this chapter we consider the decanting, in Grothendieck's terminology, of the notion of effective computability in the 1930s in the hands of the Princeton logicians on the one hand (namely Gödel, Church, Kleene and Rosser),

---

[11] See [286]. In computability theory the term "confluence" seems to be preferred over "robustness". See e.g. Robin Gandy's [77].

[12] in set theory.

[13] This is because the concept of "natural number" is usually defined in terms of finiteness. D.A. Martin expressed a similar thought in his recent [172]:

> There are various ways in which we can explain to one another the concept of the sequence of all the natural numbers or, more generally, the concept of an ω-sequence. Often these explanations involve metaphors: counting forever; an endless row of telephone poles (or cellphone towers); etc. If we want to avoid metaphor, we can talk of an unending sequence or of an infinite sequence. If we wish not to pack so much into the word "sequence," then we can say that that an ω-sequence consists of some objects ordered so that there is no last one and so that each of them has only finitely many predecessors. This explanation makes the word "finite" do the main work. We can shift the main work from one word to another, but somewhere we will use a word that we do not explicitly define or define only in terms of other words in the circle. One might worry – and in the past many did worry – that all these concepts are incoherent or at least vague and perhaps non-objective.

[14] [97], pp. 187–188.

and Alan Turing on the other. In particular, we will set what one might call, roughly, the logical approach – an approach which led (and can lead now) to foundational formalism, but which can also be pursued opportunistically and pragmatically, i.e. in the absence of foundational commitments – alongside the more semantically oriented or, as we have called it, formalism free point of view.[15] In particular we will lay the groundwork here for our review of the impact of Turing's formalism free conception of computability on the ideas of Gödel's 1946 Princeton Bicentennial Lecture in the next.

Gödel viewed Turing's analysis of computability as paradigmatic, both as conceptual analysis and as mathematical model, and the effect on his thinking was substantial.[16] Mathematically, Gödel's "transfer" of the Turing analysis of computability to the case of provability (see below) led to the first formulation of what has come to be known as Gödel's programme for large cardinals. In the case of *definability* this transfer, as one may call it, led to the fruitful concept of ordinal definability in set theory.

Philosophically much of Gödel's work (from the mid-1940s onwards) was aimed at formulating a position from which the unrestricted application of the Law of Excluded Middle to the entire cumulative hierarchy of sets could be justified. The project was intertwined with Gödel's Platonism, but it was a goal Gödel shared also with Hilbert, if not necessarily with the so-called Hilbert programme (in proof theory). Gödel's *appropriation* of the Turing analysis lent power and plausibility to his search for a logically autonomous perspective,[17] allowing an overview of logical frameworks, while not being entangled in any particular one of them – *for that is what absolute decidability entails*.

As with the notion of "finite set", one cannot think about computability, in its historical context or otherwise, without coming up against the intriguing conceptual anomalies, circularities and the like, which plagued the early attempts to ground the notion of effectivity, and plague them still: on the one hand, computation can be seen as a form of deduction; while on the other

---

[15] Foundational formalism was coined in our [124] in order to refer to the idea prevalent in foundations of mathematics in the early part of the twentieth century, associated mainly with the Hilbert programme, of embedding the mathematical corpus into a logical calculus consisting of a formal language, an exact proof concept, and (later) an exact semantics, such that the proof concept is sound and complete with respect to the associated semantics as well as syntactically complete in the sense that all propositions that can be written in the formalism are also decided.

[16] Some of Gödel's remarks on Turing computability can be found in Gödel's *193?*; see also the Gibbs Lecture, *1951* in [87], the 1965 addenda to Gödel's 1934 Princeton lectures in [47], Gödel's remarks to Wang in [280], and Gödel's correspondence. See also [47], and see below.

[17] A more general concept of autonomy is treated in Curtis Franks's [74].

hand deduction is easily seen as a form of (non-deterministic) computation. There is also the spectre of deviant encodings, which means that there is no principled way to recognize – in an absolute sense – the legitimacy of one's computational model.[18] And while we will focus on Gödel's place in these developments, the difficult question how to rule out deviant encodings is part of a more general point: semantic concepts cannot be eliminated from our meta-mathematical discourse – even (perhaps paradoxically) in the case of the concept "mechanical procedure".

3.2 Different Notions of Computability Emerge in the 1930s

Many detailed accounts have been given of the emergence of computability in the 1930s, so we do not assume the task of recounting that complex history here. We rather refer the reader to Sieg's [236], [235], [234] and [233], to Gandy's [77], to Soare's [242], to Davis's [46] and to Kleene's [134]. What follows is a condensed history of these events from our point of view, in which we follow the emergence, from the logical, of various mathematical models of human effective computability, culminating with Turing's.

In brief, then: various lines of thought ran in parallel. Gödel gave the exact definition of the class of primitive recursive functions in his landmark 1931 paper [79];[19] while in the early 1930s Church had developed the λ-calculus together with Kleene, a type-free and indeed, in Gandy's words, logic free[20] model of effective computability, based on the primitives "function" and "iteration". The phrase "logic free" is applicable only from the point of view of the later 1936 presentation of it, as Church's original presentation of the λ-calculus in his 1932 [35] embeds those primitives in a *deductive formalism*, in the Hilbert and Bernays terminology.

Church's original presentation of the λ-calculus was found by Kleene and Rosser to be inconsistent in 1934,[21] which led to Church's subsequent logic-free presentation of it, or so Kleene would imply in his history of computability in the period 1931–1933, [134]:

---

[18] Or so some have argued. See, e.g. Rescorla's [206] and the rebuttal to it by Copeland and Proudfoot in their [42].

[19] Gödel used the term "recursive" for what are now called the primitive recursive functions. The primitive recursive functions were known earlier. Ackermann [1928] produced a function which is *general recursive*, in the terminology of Gödel's 1934 Princeton lectures, but not primitive recursive. See [1]. See also Péter, [190] and [191].

[20] [77], section 14.8. As mentioned above, following Gandy we use the term "effectively computable", or just "effective", to mean "intuitively computable".

[21] About which Martin Davis said, unimprovably, "Not exactly what one dreams of having one's graduate students do for one." [77], p. 70.

When it began to appear that the full system is inconsistent, Church spoke out on the significance of λ-definability, abstracted from any formal system of logic, as a notion of number theory.[22]

Church's Thesis dates to his suggestion in 1934, to identify the λ-definable functions with the effectively computable ones. The suggestion would be lent substantial (if not as far as complete) plausibility when Church, Kleene and Rosser together proved the equivalence of λ-definability with computability in the sense of the Herbrand–Gödel equational calculus, also in 1935.[23]

Gödel, when told of Church's suggestion to equate human effective calculability with λ-definability in early 1934, found the proposal "thoroughly unsatisfactory".[24] In a letter to Kleene, Church described Gödel's then suggestion to take a logical approach to the problem:

His [Gödel's] only idea at the time was that it might be possible, in terms of effective calculability as an undefined notion, to state a set of axioms which would embody the generally accepted properties of this notion, and to do something on that basis. Evidently it occurred to him later that Herbrand's definition of recursiveness, which has no regard to effective calculability, could be modified in the direction of effective calculability, and he made this proposal in his lectures. At that time he did specifically raise the question of the connection between recursiveness in this new sense and effective calculability, but said he did not think that the two ideas could be satisfactorily identified "except heuristically."[25]

We will return to Gödel's suggestion later. For the present we note that in Church's lecture on what came to be known as "Church's Thesis" to the American Mathematical Society in 1935, Church used the Herbrand–Gödel equational calculus as a model of effective computation, i.e. recursiveness in the "new sense", rather than the λ-calculus. Perhaps he was swayed by Gödel's negative view of the λ-calculus as a model of effective computability.[26]

---

22 See their 1935 [136], which relies on Church's [38].

23 See below. The proof of the equivalence developed in stages. See Davis, [46] and Sieg, [233].

24 Church, letter to Kleene 29 November 1935. Quoted in Sieg, [233], and in Davis [46].

25 Church, [233].

26 The phrase "Church's Thesis" was coined by Kleene in 1943. See his [132]. About the attitude of Church toward the idea of taking the λ-calculus as canonical at this point Davis remarks, "The wording [of Church's published abstract JK] leaves the impression that in the early spring of 1935 Church was not yet certain that λ-definability and Herbrand–Gödel general recursiveness were equivalent. (This despite Church's letter of November 1935 in which he reported that in the spring of 1934 he had offered to Gödel to prove that 'any definition of effective calculability which seemed even partially satisfactory ... was included in λ-definability.')" Davis, [46], p. 10. See Church's 1935 abstract, [36], and Church 1936, [37].

In fact Church presented two approaches to computability in the AMS lectures and in his subsequent 1936 [37], based on the lectures: Firstly *algorithmic*, based still on what is now known as the untyped λ-calculus, i.e. the evaluation of the value fm of a function by the step-by-step application of an algorithm – and secondly *logical*, based on the idea of calculability in a logic:

> And let us call a function F of one positive integer *calculable within* the logic if there exists an expression f in the logic such that $f(\mu) = \nu$ is a theorem when and only when $F(m) = n$ is true, μ and ν being the expressions which stand for the positive integers m and n.[27]

As an aside, understanding computation as a type of mathematical derivation is very natural. It is the idea behind the Curry–Howard isomorphism, in which computability and provability are in a precise sense identified; and it has also been recently put forward by Kripke:

> My main point is this: computation is a special form of mathematical argument. One is given a set of instructions, and the steps in the computation are supposed to follow – follow deductively – from the instructions as given.[28]
>
> In particular, the conclusion of the argument follows from the instructions as given and perhaps some well-known and not explicitly stated mathematical premises. I will assume that the computation is a deductive argument from a finite number of instructions, in analogy to Turing's emphasis on our finite capacity. It is in this sense, namely that I am regarding computation as a special form of deduction, that I am saying I am advocating a logical orientation to the problem.[29]

Viewing computability in terms of logical calculi involves first restricting the class of formal systems in which the computable functions are to be represented. From Church's perspective of the 1930s, the condition in question was (essentially) that the theorems of the formal system should be recursively enumerable. Recursive enumerability is guaranteed here by two things, the so-called step-by-step argument: if each step is recursive then f will be recursive; and three conditions: (i) each rule must be a recursive operation, (ii) the set of rules and axioms must be recursively enumerable, (iii) the relation between a positive integer and the expression which stands for it must be recursive.[30]

---

[27] [37], p. 357.

[28] [144], p. 81.

[29] [144], p. 80. Kripke's point in the paper is that such arguments, being valid arguments, can be, via Hilbert's thesis, stated in a first order language. But then the solution of the Entscheidungsproblem follows almost trivially from the Completeness Theorem for first order logic.

[30] Conditions (i)–(iii) are Sieg's formulation of Church's conditions. See Sieg, [233], p. 165. For Gandy's formulation of the step-by-step argument, see [77], p. 77.

Church remarks that in imposing the restriction on the formal systems in question, he

> is here indebted to Gödel, who, in his 1934 lectures already referred to, proposed substantially these conditions, but in terms of the more restricted notion of recursiveness [i.e. primitive recursive JK] which he had employed in 1931, and using the condition that the relation of immediate consequence be recursive instead of the present conditions on the rules of procedure.[31]

We will take up Gödel's 1934 lectures below. Church's step-by-step argument effects a reduction of the notion of effectively computable function to that of calculability in a formal system of the kind, that it both satisfies conditions (i)–(iii), and the Herbrand–Gödel equational calculus has been embedded in it. A number-theoretic function is effective, in other words, if its values can be computed in a formalism which is effectively given in this sense. The argument appears to be circular.[32] Hilbert and Bernays sharpen Gödel's original conditions in their 1939 *Grundlagen der Mathematik II*, in which they present, like Church, a logical calculus rather than a system of the type of Gödel's [1934]. The essential requirement of Hilbert and Bernays is that the proof predicate of the logic is primitive recursive. This effects a precise gain: one reduces effectivity now to primitive recursion.[33]

In fact, if only because of condition (iii) of Church's argument, namely that the relation between a positive integer and the expression which stands for it must be effective, strictly speaking, the presentation in Hilbert–Bernays [1939] does not solve the circularity problem either. *The plain fact is that any analysis of effectivity given in terms of calculability in a logic, which is itself effectively given, will be subject to the charge of circularity (or, more precisely, infinite regress).* For if effectivity is explained via a logic which is itself given effectively, one must then introduce a new logic, by means of which the effectivity of the first logic is to be analysed. It is in keeping with the analysis to assume that the new logic must also be given effectively. But then one needs to introduce a third logic in terms of which the new logic is to be analysed... and so forth.[34]

---

31 [37], footnote 21, pp. 357–358.

32 See also Sieg's discussion of the "semi-circularity" of the step-by-step argument in his [233].

33 The comparison of Church [1936] with Hilbert and Bernays's [1939] follows that of Sieg's in his account [235]. See also Gandy, [77].

34 As Mundici and Sieg wrote in their [184], "The analysis of Hilbert and Bernays revealed also clearly the 'stumbling block' all these analyses encountered: they tried to characterise the elementary nature of steps in calculations, but could not do so without recurring to recursiveness (Church), primitive recursiveness (Hilbert and Bernays), or to very specific rules (Gödel)."

In his history of the period Gandy notes that a shift in perspective had set in by 1934 (Church [1936] notwithstanding) due, perhaps, to misgivings of this kind. As he writes, "in 1934 the interest of the group shifted from systems of logic to the λ-calculus and certain mild extensions of it: the $\lambda - \kappa$ and the $\lambda - \delta$ calculi".[35] Indeed the Herbrand–Gödel equational calculus, like the λ-calculus in the 1936 formulation, is also not a system of logic per se. Nor is Kleene's system presented in his 1936 [131], based on the concept of μ-recursion, a logic; and nor is Post's model of computability presented (also) in 1936 (though based on work he had done in the 1920s).[36] All of these are conceptions of computability given, primarily, mathematically – *but there was no reason whatsoever to believe in their adequacy*.

We now focus on Gödel's role in the development of computability.

3.3 The "Scope Problem"

Gödel was among the first to suggest the problem of isolating the concept of effective computability.[37] His interest in the question was driven, at least in part, by the need to give a precise definition of the notion of "formal system" – an important piece of unfinished business as far as the Incompleteness Theorems are concerned, in that it was not clear at the time to which formal systems the theorems apply, outside of *Principia Mathematica*. (We will call this the "scope problem" henceforth.)

As Gödel would realise almost immediately upon proving the Incompleteness Theorems, the solution of the scope problem depends upon the availability of a precise and adequate notion of effective computability. This is because the formal systems at issue in the Incompleteness Theorems, are to be given *effectively*. As Shapiro put the point in his [217]:

> It is natural to conjecture that Gödel's methods [in the Incompleteness Theorems JK] can be applied to any deductive system acceptable for the Hilbert program. If it is assumed that any legitimate deductive system must be effective (i.e., its axioms and rules of inference must be computable), the conjecture would follow from a thesis that no effective deductive system is complete, provided only that it is ω-consistent and sufficient for arithmetic. But this is a statement about all

---

[35] [77], p. 71.

[36] See Post's [195]. Post [195] is reprinted in Davis [47]. For a penetrating analysis of Post's work of the 1920s see de Mol, [48]. Post was apparently aware of the work of the Princeton group, but he was unaware of Turing's. See Gandy, [77].

[37] See Gandy's [77], p. 72.

computable functions, and requires a general notion of computability to be resolved.

Indeed, Gödel was careful not to claim complete generality for the Second Incompleteness Theorem in his 1931 paper:

For this [formalist JK] viewpoint presupposes only the existence of a consistency proof in which nothing but finitary means of proof is used, and it is conceivable that there exist finitary proofs that cannot be expressed in the formalism of P (or of M and A).[38]

Gödel's correspondence with von Neumann in the months following Gödel's verbal reference to the First Incompleteness Theorem in Königsberg in September of 1930, an occasion at which von Neuman was present, reveals a sharp disagreement between the two on the matter. Essentially, von Neumann saw no problem with formalising "intuitionism", whereas Gödel expressed doubts. As von Neumann wrote to Gödel the following November, "I believe that every intuitionistic [i.e. finitistic JK] consideration can be formally copied, because the 'arbitrarily nested' recursions of Bernays-Hilbert are equivalent to ordinary transfinite recursions up to appropriate ordinals of the second number class".[39] And in January 1931 von Neumann responded to the above-cited disclaimer of Gödel's, having seen the galleys of Gödel's 1931 paper:

I absolutely disagree with your view on the formalizability of intuitionism. Certainly, for every formal system there is, as you proved, another formal one that is ... stronger. But intuitionism is not affected by that at all.[40]

Gödel responded to von Neumann, "viewing as questionable the claim that the totality of all intuitionistically correct proofs are contained in *one* formal system".[41] And as he wrote to Herbrand in 1931:

Clearly, I do not claim either that it is certain that some finitist proofs are not formalizable in Principia Mathematica, even though intuitively I tend toward this assumption. In any case, a finitist proof not formalizable in *Principia Mathematica* would have to be quite extraordinarily complicated, and on this purely practical

[38] [84], p. 195. P is a variant of *Principia Mathematica*.
[39] [89], p. 339.
[40] [89], p. 341.
[41] Gödel responded to von Neumann in writing, but his letters seem to have been lost. See [236], p. 548. We know of his response through the minutes of a meeting of the Schlick Circle that took place on 15 January 1931, which are found in the Carnap Archives of the University of Pittsburgh. See Sieg's introduction to the von Neumann–Gödel correspondence in [89], p. 331.

ground there is very little prospect of finding one; but that, in my opinion, does not alter anything about the possibility in principle.[42]

By the time of Gödel's 1933 Cambridge lecture Gödel seems to have reversed himself on the question: "So it seems that not even classical arithmetic can be proved to be non-contradictory by the methods of the system *A* ..."[43] Nevertheless Herbrand and Gödel had agreed in their correspondence that the concept of finite computation was itself "undefinable", a view Gödel held through 1934 (and beyond), when he wrote the oft-quoted footnote 3 to the lecture notes of his Princeton lectures:

> The converse seems to be true if, besides recursions according to the scheme (2), recursions of other forms (e.g. with respect to two variables simultaneously) are admitted. This cannot be proved, since the notion of finite computation is not defined, but serves as a heuristic principle.[44]

Looking ahead, Gödel took later a radically different view of the matter. As he wrote in the amendment to footnote 3 in the 1965 reprinting of his 1934 lectures: "This statement is now outdated; see the Postscriptum, pp. 369–71."[45] In the Postscriptum Gödel indicates that the Turing analysis gave a completely adequate analysis of the concept of finite computation. The Turing analysis would not settle the general question, of course, of the adequacy of any given formalism for the concept of informal provability *überhaupt*. As we will see in the next chapter, in his 1946 Princeton Bicentennial Lecture Gödel would take a negative view of the matter, expressing the opinion that no single formalism is adequate for expressing the general notion of proof – so not just the notion of finite proof. In fact the view is already expressed in Gödel's 1933 Cambridge lecture: "So we are confronted with a strange situation. We set out to find a formal system for mathematics and instead of that found an infinity of systems ..."[46]

Returning to the scope problem, although the Second Incompleteness Theorem aims at demonstrating the impossibility of giving a finitary consistency proof in the systems considered, this does not settle the general problem to which formal systems the Incompleteness Theorems apply. As for the problem of isolating the mathematical concept of effectivity, as Sieg notes of the

---

[42] [88], p. 23.
[43] [87], p. 51. See Feferman's discussion of this point in his introduction to *1933o*.
[44] The claim the converse of which is being considered here, is the claim that functions computable by a finite procedure are recursive in the sense given in the lectures. [84], p. 348.
[45] Gödel's addenda to the 1934 lectures were published in Davis's [47].
[46] [87], p. 47.

Herbrand correspondence, "Nowhere in the correspondence does the issue of general computability arise."[47]

By 1934, compelled to "make the incompleteness results less dependent on particular formalisms",[48] and somewhat at variance with the axiomatic approach he had suggested to Church earlier, Gödel introduced, in his Princeton lectures, the general recursive, or Herbrand–Gödel recursive functions, as they came to be known, defining the notion of "formal system" as consisting of "symbols and *mechanical rules* relating to them".[49] Both inference and axiomhood were to be verified by a finite procedure:

> ... for each rule of inference there shall be a finite procedure for determining whether a given formula B is an immediate consequence (by that rule) of given formulas $A_1, \ldots A_n$, and there shall be a finite procedure for determining whether a given formula A is a meaningful formula or an axiom.[50]

The Herbrand-Gödel recursive functions are mathematically rather than logically presented. As Gödel remarked in the opening lines of the lecture in connection with the function class, these are "considerations which for the moment have nothing to do with a formal system".[51]

The calculus admits forms of recursions that go beyond primitive recursion. Roughly speaking, while primitive recursion is based on the successor function, in the Herbrand–Gödel equational calculus one is allowed to substitute other recursive functions in the equations, as long as this defines a unique function. (Since for example $f(n) = f(n+1)$ does not define a unique function.) It was not clear to Gödel at the time, and prima facie it is not clear now, that the schema captures *all* recursions. As Gödel would later write to Martin Davis, "I was, at the time of these lectures, not at all convinced that my concept of recursion comprises all possible recursions."[52]

Leaving the adequacy question aside for the moment, the equational calculus effects a reduction of one conceptual domain – the strictly

---

[47] [234], p. 180.

[48] Sieg [236], p. 554.

[49] [84], p. 349. Emphasis added. Such a mechanistic view of the concept of formal system was not a complete novelty at the time. Tarski conceived of "deductive theory" for example, as "something to be performed". See Hodges, [106]. As Hodges put it, Tarski's view was "that a deductive theory is a kind of activity".

[50] [84], p. 346.

[51] [84], p. 346. This contrasts with Church's initial presentation of the λ-calculus.

[52] Quoted in [84], p. 341. And as Kleene would later write, "Turing's computability is intrinsically persuasive but λ-definability is not intrinsically persuasive and general recursiveness scarcely so (its author Gödel being at the time not at all persuaded)." See Kleene's 1981 [135].

proof-theoretic – to another, the mechanical. Gödel's shift of gaze, as it were, was in keeping with the shift in orientation of the Princeton group at the time. The shift may have been driven by the scope problem – for how else to make the concept of formal system exact? Analysing the concept of "formal system" in terms of the concept of "formal system" itself, would have been no analysis at all.

Returning to Gödel's 1934 lectures, Gödel gives later in the paper the precise definition of the conditions a formal system must satisfy so that the arguments for the incompleteness theorems apply to it. In addition to conditions involving representability and Church's condition (iii), the relevant restriction Gödel imposed was

> Supposing the symbols and formulas to be numbered in a manner similar to that used for the particular system considered above, then the class of axioms and the relation of immediate consequence shall be [primitive JK] recursive.[53]

This attaches a primitive recursive characteristic function to each rule of inference and gives a reduction of the concept of "formal system" to recursivity in the sense of the Herbrand–Gödel equational calculus.[54]

As we observed of the analysis of computability Church gave in his [37], this lends an air of circularity to the analysis (of the notion of formal system). But equally pressing to Gödel may have been the identification of formal provability with computability in the sense of the Herbrand–Gödel equational calculus. Prima facie, there is no compelling reason to suppose even the extensional equivalence of these notions.

We permit ourselves a brief digression in order to mention Gödel's presentation in these lectures of the theorem expressing the undecidable sentence from his 1931 paper in diophantine form. This theorem expressing the self-referential statement in the form of a polynomial equation remains underappreciated by working mathematicians, if not completely unknown, even today. The theorem resonates with the approach Post had laid out in his 1944 [196], and it has an interesting history: at the Königsberg meeting in 1930, in private discussion after the session at which Gödel announced the First Incompleteness Theorem, von Neumann asked Gödel whether the undecidable statement in question could be put in "number-theoretic form", given the fact of arithmetisation. Gödel was initially sceptical but eventually proved the diophantine version of the theorem, to his surprise.[55]

---

53 [84], p. 361.
54 See also Sieg, [236], p. 551.
55 See Wang, [279]. Tarski in his 1946 Princeton Bicentennial Address pressed the issue further, asking for undecidable statements coming from "ordinary" mathematics. See Sinaceur [237].

The proof of the equivalence between the diophantine and the explicitly self-referential statement of Gödel's 1931 paper turns on the fact that the Δ_0 subformula which occurs in the undecidable Π_1 sentence has a recursive characteristic function. We stress again that here, "recursive" would have meant in the sense of Herbrand–Gödel equational calculus, a fact which may have weakened claims to generality also in that case.

By 1935, Gödel's reflections on computability in the higher order context began to point toward the possibility of a definitive notion of formal system. Nevertheless, an, in Gödel's terminology now, *absolute* definition of effective computability was still missing at that point. As Gödel wrote to Kreisel in 1965:

> That my [incompleteness] results were valid for all possible formal systems began to be plausible for me (that is since 1935) only because of the Remark printed on p. 83 of 'The Undecidable' ... But I was completely convinced only by Turing's paper.[56]

Gödel's "Remark" is contained in the addendum to the 1936 abstract "On the Lengths of Proofs":

> It can, moreover, be shown that a function computable in one of the [higher order JK] systems S_i, or even in a system of transfinite order, is computable already in S_1. Thus the notion "computable" is *in a certain sense* "absolute", while almost all metamathematical notions otherwise known (for example, provable, definable, and so on) quite essentially depend on the system adopted.[57]

The above passage seems to be the first hint in Gödel's writings of preoccupations that would materialise more fully in his 1946 Princeton lecture. The fact that the passage from lower to higher order quantification does not give any new recursive functions is easily seen once Kleene's Normal Form Theorem is in place. Kleene's theorem exhibits a universal predicate, in which a large numerical parameter, which codes the entire computation, is "guessed". The theorem played a crucial role in the acceptance of Church's Thesis, both on Gödel's part, in combination with Turing's work, and generally. This is because Kleene Normal Form is the way to establish *confluence*, the equivalence of different mathematical notions of computability.[58]

---

[56] Quoted in Sieg [235], in turn quoting from an unpublished manuscript of Odifreddi, p. 65.

[57] "On the length of proofs," [84],
p. 399. Emphasis added.

[58] See Davis's [46] and see below. Kleene's Normal Form Theorem is stated in [46] as follows: Every general recursive function can be expressed in the form $f(\mu y(g(x_1 .....x_n, y) = 0))$, where f and g are primitive recursive functions.

As for Gödel's view of the adequacy of the models of computation which had emerged at the time, his premature departure from Princeton in the spring of 1934 put him at some remove from subsequent developments. In particular he would not have been a direct witness to the further empirical work Church, Kleene and Rosser had done, showing that the known effective functions were λ-definable.[59] Gödel had taken ill and would not be on his feet for some time; and he had now turned to the continuum problem.

3.4 Turing's Analysis of Computability

In 1936, unbeknownst to the logicians in Princeton, Turing gave a self-standing analysis of human effective computability and used it to solve the *Entscheidungsproblem*, as Church had solved it just prior.

Rather than calculability in a logic, Turing analysed effectivity informally but exactly, via the concept of a Turing Machine – a machine model of computability, consisting of a tape scanned by a reader, together with a set of simple instructions in the form of quadruples.[60] More precisely, the analysis consisted of two elements: a conceptual analysis of human effective computation, together with a mathematical precisification of it consisting of rules given by a set of quadruples, as follows: "erase", "print 1", "move left", and "move right".

We alluded to circularity in connection with approaches to computability centred on the idea of calculability in a logic.[61] The crucial point here is that the Turing analysis of the notion "humanly effectively computable" does not involve the specification of a logic *at all*. Of course, if one defines a formalism, or alternatively a *logic*, as we have done here then with very little mathematical work *each of the informal notions of computability we have considered so far can be seen as generating formal calculi in this sense*, i.e. as an individual, self-standing formalism with its own syntax and rules of proof and so forth. In that sense the concept of a Turing Machine may *give rise* to a logic; or have a logic embedded within it in some very generalised sense. But a logic it is not.[62]

---

[59] This point was made by Dana Scott in private communication with the author.

[60] Or quintuples, in Turing's original presentation.

[61] By a *logic*, we have meant a combination of a list of symbols, commonly called a signature, or vocabulary; rules for building terms and formulas, a list of axioms and rules of proof, and then, usually, a semantics.

[62] See also Kanamori, "Aspect-perception and the history of mathematics", [115].

The reaction to Turing's work among the Princeton logicians was immediately positive. As Kleene would write in 1981, "Turing's computability is intrinsically persuasive but λ-definability is not intrinsically persuasive and general recursiveness scarcely so (its author Gödel being at the time not at all persuaded)." As Gödel would later explain to Hao Wang, Turing's model of human effective calculability is, in some sense, perfect:

> The resulting definition of the concept of mechanical by the sharp concept of "performable by a Turing machine" is both correct and unique ... *Moreover it is absolutely impossible that anybody who understands the question and knows Turing's definition should decide for a different concept.*[63]

> The sharp concept is there all along, only we did not perceive it clearly at first. This is similar to our perception of an animal far away and then nearby. We had not perceived the sharp concept of mechanical procedure sharply before Turing, who brought us to the right perspective.[64]

Gödel would use the language of "Kantian ideas" to conceptualize this process of sharpening:

> Absolute demonstrability and definability are not concepts but inexhaustible [Kantian] ideas. We can never describe an Idea in words exhaustively or completely clearly. But we also perceive it, more and more clearly. This process may be uniquely determined – ruling out branchings.[65]

We will return to Gödel's notion of absolute demonstrability and definability as Kantian Ideas in the next chapter. Returning to the scope problem, Turing's conceptual analysis led to its complete solution in Gödel's view. We cited Gödel's 1965 letter to Kreisel above; other evidence to this in the record is to be found in the 1965 publication of the notes of Gödel's 1934 Princeton lectures, this time with new footnotes, and including the following Postscriptum:

> In consequence of later advances, in particular of the fact that, due to A. M. Turing's work, a precise and unquestionably adequate definition of the general concept of formal system can now be given, the existence of undecidable arithmetical propositions and the non-demonstrability of the consistency of a system in the same system can now be proved rigorously for every consistent formal system containing a certain amount of finitary number theory.

---

[63] Remark to Hao Wang, [280], p. 203. Emphasis ours.
[64] ibid, p. 205.
[65] ibid, p. 269.

... Turing's work gives an analysis of the concept of "mechanical procedure" (alias algorithm or computation procedure or "finite combinatorial procedure"). This concept is shown to be equivalent with that of a "Turing machine." A formal system can simply be defined to be any mechanical procedure for producing formulas, called provable formulas. For any formal system in this sense there exists one in the [usual] sense that has the same provable formulas (and likewise vice versa) ... [66]

Beyond what he said above, Gödel would not explain his endorsement of Turing's machine model at any length in his writings, but rather, simply, point to it from time to time, as the only real example of a fully worked out conceptual (even phenomenological) analysis in mathematics. The "epistemic work", so to speak, that is, the sharpening of intuition usually brought out by formalisation, was brought out this time by an informal analysis. *And in fact the analysis had to go this way.* One cannot ground the notion of "formal system" in terms of concepts that themselves require the specification of this or that formal system.

Wang explains that Turing's analysis had even ontological significance for Gödel, in that "Turing machines are an important piece of evidence for Gödel's belief that sharp concepts exist and that we are capable of perceiving them clearly".[67]

The sharp concept is there all along, only we did not perceive it clearly at first. [68]

The concept of a Turing Machine is formalism free, in our sense of the term, while at the same time it is a concept from which the formal notion can be easily recovered. A Turing Machine is neither a formal nor an informal concept, however; or, one could say, it is, in some strange sense, both[69] – the *one thing*, the missing piece which anchors the picture.

Looking ahead, this correspondence between the informal notion of computability and the various formal notions that had been introduced to date, effectuated via a "perfect" analysis of the informal concept, a correspondence Robin Gandy described as a theorem in his [77], is what Gödel is asking for in his 1946 lecture – not as applied to computability, as Turing had already done that, in Gödel's view. What is needed is to transfer the entire Turing analysis to the "other" epistemological cases, namely definability and provability.

---

[66] [84], p. 369.
[67] ibid, p. 194.
[68] ibid, p. 205.
[69] As Juliet Floyd puts it in [68], "A Turing machine lends itself, intentionally and conceptually, to a double point of view: it is both a formal system and a remodeling."

As for the Turing Machine, for the logicians of the time, and for many logicians today, the Turing Machine is not just another in the list of acceptable notions of computability – it is the *grounding* of all of them. Turing held a mirror up to a specific human act: a computor following a fixed routine. The human profile is carried right through to the end, smuggled into mathematical terrain *nec plus, nec minus, nec aliter*. It is as seamless a fit of the raw and the cooked as ever there was in mathematics, and there is now, of course, no problem of adequacy. It was "the focus on the user, the human end"[70] that makes the construction so indubitably *right*. The Turing Machine is the beating heart of computability theory.

Gandy saw Turing's isolation from the logical milieu of Princeton as the key to his discoveries:

> It is almost true to say that Turing succeeded in his analysis because he was not familiar with the work of others ... The bare hands, do-it-yourself approach does lead to clumsiness and error. But the way in which he uses concrete objects such as exercise book and printer's ink to illustrate and control the argument is typical of his insight and originality. Let us praise the uncluttered mind.[71]

> All the work described in Sections 14.3–14.9[72] was based on the mathematical and logical (and *not* on the computational) experience of the time. What Turing did, by his analysis of the processes and limitations of calculations of human beings, was to clear away, with a single stroke of his broom, this dependence on contemporary experience, and produce a characterization which – within clearly perceived limits – will stand for all time.[73]

What was the role of confluence? Church's Thesis, as originally suggested by Church in 1934, identified effective calculability with λ-calculability, and then with Herbrand–Gödel calculability, after the equivalence between the two was established. Confluence is established in a weak metatheory by means of the Kleene Normal Form Theorem, also in the case of proving the equivalence of Turing computability with the remaining notions. Prior to Turing's work, the available confluence was seen as only weakly justifying adequacy. Once one has a grounding example in hand this changes – confluence now plays an epistemologically important evidentiary role.

3.5 Gödel's Reaction to Turing's Work at the Time

How soon after learning of Turing's work would Gödel see the Turing analysis in such vigorously positive terms as are characteristic of his later

---

[70] Floyd, [68].
[71] [77], p. 83.
[72] i.e. the work on computability of Church and others prior to Turing's 1937 paper
[73] ibid, p. 101.

communications with Kreisel and Wang, and of his 1965 writings? The text *193?*,[74] dating presumably from the years 1936–39, is informative on this point. In the text, Gödel gives a perspicuous presentation of the Herbrand–Gödel equational calculus. He also improves the result of the 1934 lectures, that undecidable sentences for the formal theories in question can be given in the form of diophantine equations, by showing that the diophantine equations in question are limited in degree and in the number of variables (while not actually computing the bound). He expressed his view of Turing's work at the time thus:

> When I first published my paper about undecidable propositions the result could not be pronounced in this generality, because for | the notions of mechanical procedure and of formal system no mathematically satisfactory definition had been given at the time. This gap has since been filled by Herbrand, Church and Turing. The essential point is to define what a procedure is. Then the notion of formal system follows easily ...

Of the version of the Herbrand–Gödel equational calculus presented here, Gödel goes on to say the following: "That this really is the correct definition of mechanical computability was established beyond any doubt by Turing." That is, using the calculus one can enumerate all "possible admissible postulates", in the terminology of the notes; but since one cannot decide which of the expressions defines a computable function, one cannot diagonalise out of the class, "And for this reason the antidiagonal sequence will not be computable."[75] Gödel then goes on to say, "But Turing has shown more ..." The "more" Turing has shown, is the equivalence of the class of functions computable by Herbrand–Gödel calculus, with the class of functions computable by a Turing Machine. Gödel seems to be saying here that the Herbrand–Gödel equational calculus earns adequacy by virtue of its equivalence with Turing's notion – by inheritance, as it were.

We saw an analytical schema at work in the development of computability in the 1930s, a schema bringing *confluence* together with *grounding*. We focussed here on Gödel's absorption of the Turing analysis at the time and over the subsequent decade, keeping an eye on the scope problem and the development, in tandem, of the notion of *formal system*. In the next chapter we will consider a shift in Gödel's work, namely the entrenchment of a methodological schema one may think of as a *generalised form of the Turing*

---

[74] [87], p. 164.
[75] [87], p. 168.

analysis. Directed outward from computability, the emphasis here is on consequence and on definability, or "comprehensibility by the mind", in the service
of *logical autonomy.*

We will see that Gödel's search for autonomy took a particular form: finding
absolute definitions of three fundamental epistemological notions: computability, definability and provability. The project imposed, *and imposes on us,*
contradictory demands: the logician must control her logic, while at the same
time seeking to free herself of it. The logician is forced to look in two directions
at once – toward mathematics done purely conceptually and in the pleasant
and comfortable vagueness of natural language; and then toward logical and
linguistic exactness.

What interested us in this chapter happened on a different, and perhaps
smaller stage: certain slight shifts of attention on Gödel's part, and on the
part of other logicians working in the 1930s; tentative, first step attempts at
autonomy, that, if one is attuned to them in the right way, gained Gödel new
logical ground.

3.6 Coda: A Word about Deviant Encodings

In a strange twist of fate the essentially *philosophical* problem of deviant
encodings, as it is now called, cannot be avoided. That is, Gödel's condition in
his 1934 Princeton lectures, that "the relation between a positive integer and
the expression which stands for it must be recursive" will *always* introduce
circularity, insofar as the condition is a constituent of an all-embracing, or in
Gödel's other terminology, absolute notion of computability.

In brief: As a general rule, Turing Machines require an input, and at the
end of the computation they generate an output. Inputs and outputs must be
encoded, as, after all, Turing Machines operate on finite strings, not natural
numbers.

There are, of course, more or less obvious ways to do this. The obvious
choice, "unary" encoding, involves counting the number of consecutive ones
in the beginning of the tape and declaring that as the input, and, respectively,
counting at the end of the computation the number of consecutive ones in the
beginning of the tape, and declaring that as the output. As far as one can see,
this approach is completely unproblematic! *But one has to accept the effectivity
of unary encoding by faith.*

In fact, there are also other, equally obvious encodings. For example, one
could represent the input number as a binary string which would then be written on the tape. This should be as computable as unary encoding, and in fact

it is even more effective because one gets logarithmic compression. But again the effectivity of the (binary) encoding must be taken on faith.

The problem of "deviant encodings", pointed out e.g. by Shapiro in his 1982 [216], is that the encoding of input and output data on the tape should be itself computable, devoid of unintended elements which may give the machine non-computable power. For example, using Copeland and Proudfoot's deviant "mirabilis" output encoding, one can design a Turing Machine that computes the Halting Problem.

Who would question the effectivity of unary encoding? Nobody we know! The problem is that one would like to have a general method for separating computable from non-computable encodings, rather than a list of acceptable ones. But such a general criterion cannot be given.[76]

---

[76] Rescorla has claimed recently in [206] that because of this problem, "we are still awaiting a genuine conceptual analysis" of computability. We do not agree, for the reasons we have given here.

4

Gödel and Formalism Independence

Is formalisation like death? Gödel recorded a thought about this in 1944, in the eleventh of his so-called "MaxPhil" notebooks:

> Remark (Philosophy): Is the difference between living and lifeless perhaps that its laws of action don't take the form of "mechanical" rules, that is, don't allow themselves to be "formalized"? That is then a higher level of complication. Those whose intuitions are opposed to vitalism claim then simply: everything living is dead.[1]

One has to treat the esoteric writings of a philosopher with care. It is clear that Gödel was somewhat bolder, philosophically, in private, than in his public writings and remarks, as one would expect.[2]

In MaxPhil notebook IX Gödel set down a milder thought about formalisation and – this time – games:

> A board game is something purely formal, but in order to play well, one must grasp the corresponding content [the opposite of combinatorically]. On the other hand the

---

[1] Gödel's MaxPhil notebooks are numbered I–XV, with volume XIII either missing or never existing. The notebooks are written in Gabelsberger shorthand and are part of the archive entitled Gödel Papers, Shelby White and Leon Levy Archives Center, Institute for Advanced Study, Princeton, NJ, USA, on deposit at Princeton University. MaxPhil XI seems to have been written in 1944. The quote is from Max XI 030097 (box 6b folder 70). Transcription from the Gabelsberger by Maria Hämeen-Anttila, translation from the German by the author. Second reading by Mark van Atten, Paola Cantu, Gabriella Crocco, Eva-Maria Engelen. The German text: "Bemerkung (Philosophie): Ist der Unterschied des Lebendigen vom Leblosen vielleicht da, dass sich seine Wirkungsgesetze nicht in "mechanischen" Regeln fassen, d.h. nicht "formalisieren" lässen? D.h. also eine höhere Stufe der Komplikation die dem Vitalismus entgegengesetzte Anschauung behauptet also einfach: alles Lebendige ist tot."

[2] In Wang's *Logical Journey* Gödel is recorded as saying: "I am cautious and only make public the less controversial parts of my philosophy." Wang, [280], remark 7.3.14. To this point Kreisel observed to the author once, that whereas Wittgenstein was a "wild" in his writing but "sensible" in conversation, Gödel was the opposite.

formalization is necessary for control [note: for only it is objectively exact], therefore knowledge is an interplay between form and content.[3]

We have here the distinction between content and form, and the idea that content lies in some sense outside of "form", a mainstay of Gödel's philosophical writings;[4] the contrast between having direct knowledge of content, as opposed to knowledge mediated by combinatorics of syntax; the role of formalisation as the only source of "objective exactness", of *control*.

In this chapter we will argue that formalism independence emerges as a valued third party in this complex world picture, drawing on evidence taken from Gödel's 1946 Princeton Bicentennial Lecture [86]. We will briefly consider other sources, e.g. some remarks in Gödel's thesis about the possibility of an informal decision procedure, as well as some of his remarks on content from his 1958 *Dialectica* article [83].[5] The importance of the *Dialectica* article for us is that it represents another step toward "logic-freeness", to borrow the terminology of A. Troelstra in his editorial introduction to the article in [86].

4.1 Gödel on Formalisation

Gödel's view of formalisation was obviously complex. At times it was simply the standard one: formalisation assists the mathematician by giving control over proofs:

> Incidentally, the best method, in my opinion, of checking a complicated proof is to bring it close to formalization by interpolating so many auxiliary theorems that the

---

[3] Gödel, Max IX 030095 (box 6b folder 69). Transcription from the Gabelsberger by Maria Hämeen-Anttila. Also quoted in Floyd and Kanamori, [69], translation from the German by J. Floyd. The text immediately preceding this in Max Phil IX is: Bem[erkung] (Phil[osophie]): Obzwar die Form etwas höheres ist als der Inhalt und (wahrscheinlich) im Wesen alle[s] Form ist, so ist doch "Verstehen" = "Inhalt sehen", und auch die Form [ist] nur dadurch verständlich, dass sie "verinhaltlicht" wird. Nicht jede, sondern nur gewisse [die fruchtbaren und objektiv existierenden] Formen haben einen entsprechenden "Inhalt" durch den sie "verstanden" werden; das heisst, dass man sie dann durch Blick und in Anwendung beherrscht. Das "Verinhaltlichen" ist in einem gewissen Sinn das Gegenteil des "Formalisierens". Jedenfalls liegt der richtige Zugang zur höchsten Form im höchsten Inhalt [Jesum[[voraus]] Maria] und sind die mat[hematischen] Wissenschaften dazu nicht geeignet. [Aber vielleicht kann man den Inhalt, der ihnen zu Grunde liegt, erfassen?] Arist[oteles] und Plato [im Gegensatz zu Leibniz und Desc[artes]] untersuchen die höchsten Inhalte [nicht Formen], d.h. Subst[anz], Kausal[ität], Sein etc. Transcription from the Gabelsberger by Maria Hämeen-Anttila.

[4] See e.g. the argument from content in the Gibbs lecture [*1951] in [87].

[5] "Über eine bisher noch nicht benützte Erweiterung des finiten Standpunktes."

proof of each becomes trivial on the basis of the preceding ones. This procedure frequently also leads to simplifications or generalizations.[6]

Here is Gödel on the notion of a *formalism* in a letter to Bernays:

> As to the book by Speiser, I find the introduction [i.e., the Präludium] very
> · beautiful, *particularly because a precise formalism is given*. But the detailed
> execution, as far as I've read it, I find not a bit clearer than Hegel himself.[7]

This is the old conception of logic and formalisation as a methodology for simplifying theorems and laying bare essential structure; the conception of logic as having the potential either to break new ground in solving problems mathematicians were otherwise unable to solve – or to give the reason for the lack of a solution. And while nowadays the mathematician may value formalisation for its negative role in their practice, e.g. for its capacity to demonstrate the limitations of the standard methods through the set-theoretic independence results,[8] one would be hard put to find a working mathematician who embraced the old conception in its full form.

Gödel adopted the old conception at various times, as we saw; but equally often Gödel expressed the idea that formalisation has not fulfilled the hopes that logicians may have had for it:

> Perhaps the reason why no progress is made in mathematics (and there are so many
> unsolved problems), is that one confines oneself to the ext[ensional] thence also the
> feeling of disappointment in the case of many theories, e.g., propositional logic and
> formalisation altogether.[9, 10]

---

[6] Letter to Boone of December 1956 in [88] p. 327. This is the "standard view of proof", as it has recently been called, namely the view that an informal proof is valid, or in other versions of the standard view, rigorous, only to the extent that it can be formalised in an accepted proof system. As S. MacLane put it:

> A mathematical proof is rigorous when it is (or could be) written out in the first order
> predicate language $L(\in)$ as a sequence of inferences from the axioms ZFC, each inference
> made according to one of the stated rules … [158], pp. 377–378.

The standard view has been challenged recently. See for example [49], [90], [147] and [205].

[7] Letter to Bernays of July 1962 in [88], p. 207. Emphasis ours.

[8] See e.g. Shelah's proof of the independence of the Whitehead Conjecture [221].

[9] Gödel, Max IV 030090, p. 198. Transcription Cheryl Dawson; translation from the German ours; amendment ours. Gödel's dating of Max IV indicates that the remark was written in the period from May 1941 to April 1942.

[10] Gödel used the term "extensional" to chastise mathematicians also in a letter to Bernays:

> [Mr. Dilljer] wrote me a letter about his work (done jointly with Nahm). I don't understand
> what it is supposed to mean that in my proof of the formula $p \rightarrow p \wedge p$ an (impossible)
> passage to the characteristic term of a formula is necessary. What is necessary is the
> decidability of intensional equations between functions. The mathematicians will probably

To work extensionally means, for Gödel here, to eschew *meaning*; more broadly, to prescind from the kind of deep conceptual analysis that one associates with the kind of philosophy Gödel admired, German Idealism and such. As for formalisation and extensional mathematics, Gödel is noticing that a formalisation, in the "blind sense", is *asemic* – it has no semantic content. In fact one might think that is its virtue. "Blind formalism" was the phrase Gödel often used, e.g. in a letter to L. Rappaport:

> I have not proved that there are mathematical questions undecidable for the human mind, but only that there is no machine (or blind formalism) that can decide all number theoretical questions (even of a certain very special kind). ... Likewise it does not follow from my theorems that there are no convincing consistency proofs for the usual mathematical formalisms, notwithstanding that such proofs must use modes of reasoning not contained in those formalisms. What is practically certain is that there are, for the classical formalisms, no conclusive combinatorial consistency proofs (such as Hilbert expected to give), i.e. no consistency proofs that use only concepts referring to finite combinations of symbols and not referring to any infinite totality of such combinations. I have published lately (see Dial., vol. 12 (1958)a p. 280) a consistency proof for number theory which probably for many mathematicians is just as convincing as would be a combinatorial consistency proof, which however uses certain abstract concepts (in the sense explained in this paper). Your formulation is overdramatized and not true as it stands. It is not the structure itself of the deductive systems which is being threatened with a break/down, but only a certain interpretation of it, namely its interpretation as a blind formalism.[11]

What would it mean, for Gödel, for a formalism *not* to be blind? And what is "the structure itself" of the deductive systems, recalling Gödel's note to himself in MaxPhil IX that "Obzwar die Form etwas höheres ist als der Inhalt und (wahrscheinlich) im Wesen alle[s] Form ist"?[12] Can a formalism (in Gödel's expanded sense of the term) have semantic content at all?

MacFarlane questioned the idea that expressing a logic in syntactic terms eliminates semantic content, as we saw in chapter 2.[13] And in this vein

---

raise objections against that, because contemporary mathematics is thoroughly extensional and hence no clear notions of intensions have been developed. (July 1970 letter to Bernays in [88], p. 283.)

[11] [89], pp. 176–177.

[12] See footnote[3].

[13] "The fact that certain logics can be formulated in completely syntactic terms, without reference to the meanings of their symbols, does not support the claim that they have no semantic content." From "What does it mean to say that logic is formal?", John Gordon MacFarlane, p. 36, Ph.D. thesis, University of Pittsburgh. The remark is part of a broader point that different notions of formality or logicality tend to be conflated in the literature.

M. Detlefsen has written that even Hilbert himself saw a role in metamathematics for *contentual* proof:

> Hilbert's "decontentualization" of proof – his proposed replacement of propositions and other contentual items which figure centrally in the traditional conception of proof by the formal objects of (his formal conception of) axiomatic proof was thus in his view a transformation that is necessary if the legitimate demands of rigor are generally to be met. His view is complicated by the fact that, in addition to urging a place for the above-described conception of axiomatic proof and its accompanying conception of rigor, Hilbert continued to see a place in mathematics (and metamathematics) for contentual proof as well.[14]

Hilbert (and his co-workers) had a clear view about formalisation and content; while Gödel expressed himself somewhat less frequently on the topic of formalisation per se. By using the modifier "blind", Gödel may allow for the possibility that a formalisation as such is not necessarily "blind", that it can have semantic content.

In a recent talk, Arana set out a distinction between formalisation as *extraction* and formalisation as *compression*. He quotes Hilbert in connection with the former notion:

> In Hilbert's *Foundations of Geometry*, the axiomatic standpoint has been sharpened regarding the axiomatic development of a theory: "From the factual and conceptual subject matter that gives rise to the basic notion of the theory, we retain only the essence [Extrakt] that is formulated in the axioms, and ignore all other content."[15] In moving from a geometric data set to a (modern) axiomatization, we strip away all, for instance, visual content, following Pasch. This is what Poincaré was expressing.[16]

---

[14] See Detlefsen's "Abstraction, axiomatization and rigor: Pasch and Hilbert" [54], p. 175. Hilbert's distinction between contentual and noncontentual inference is described in many places, for example in his "On the infinite" [102]:
We now divest the logical signs of all meaning, just as we did the mathematical ones, and declare that the formulas of the logical calculus do not mean anything in themselves ... In a way that exactly corresponds to the transition from contentual number theory to formal algebra we regard the signs and operation symbols of the logical calculus as detached from their contentual meaning. In this way we now finally obtain, in place of the contentual mathematical science that is communicated by means of ordinary language, an inventory of formulas that are formed from mathematical and logical signs and follow each other according to definite rules. Certain of these formulas correspond to the mathematical axioms, and to contentual inference there correspond the rules according to which the formulas follow each other; hence contentual inference is replaced by manipulation of signs according to rules, and in this way the full transition from a naïve to a formal treatment is now accomplished. (p. 381)

[15] Arana cites Hilbert & Bernays, Grundlagen der Mathematik, Vol. 1, 1934 [101].

[16] Author meets Critics session on J. Baldwin's [10] CLMPS Prague 2019.

Arana contrasts this notion of formalisation, so keeping "axiomatic content" but eliminating all other content, with the notion coming out of art, in which formalisation (in the form of abstraction) *compresses* meaning, but does not eliminate it. Arana quotes the art historian Wilhelm Worringer:

> [Abstraction is] to wrest the object of the external world out of its natural context, out of the unending flux of being, to purify it of all its dependence upon life, i.e. of everything about it that was arbitrary, to render it necessary and irrefragable, to approximate it to its absolute value.[17]

This is abstraction as *compression*, or purification. As an aside, the art historian Briony Fer notes that a similar set of contrasts emerged early on, and at more or less the same time, in the various camps that had formed around (formalisation in the form of) abstract painting.[18] Thus she cites the art historian Carl Einstein as an exemplar of resistance to certain abstract painters' "mathematical drunkenness", likening its repetition of forms to "police methods", derived, in a remark slightly reminiscent of Gödel's,[19] "from the anxiety before the invisible and before the sudden disappearance by death". Like Worringer, Einstein was signalling toward a different, noneliminativist, view of abstraction: that of the signifier, and the field of play, of certain psychological forces:

> What Einstein saw was not a logic at work in the form of the square, but some lurching, garbled undertow dragging it back to a set of unconscious mechanisms.[20]

The task that preoccupies Fer in [64] is not to chronicle the origins of this polarity of abstractions, or, looking forward, to track its course through modernity; her interest is rather "in the impossibility of keeping these twin poles apart for long".[21] This is our interest too – for the categories of syntax and semantics. We will return to Fer's remark in the last chapter.

A common view of formalisation is evinced in Putnam's 1980 "Models and reality" [200], in which Putnam entertained the idea of a first order formalisation of our best scientific theories, and beyond, a formalisation of "our entire body of belief". From the autonomy of natural language point of view put forward in this book, the idea is out of the question, on the grounds that no such

---

[17] *Abstraction and Empathy*, 1908.
[18] Fer, *On Abstract Art* [64], pp. 2–4.
[19] on formalisation and death
[20] Fer, ibid.
[21] Fer, ibid.

theory is available which would be sufficiently meaning-preserving – provided that such a theory is possible at all.[22]

Kreisel seems to have had unresolvedly complicated views about formalisation, as we saw. Logical hygiene is not "fundamental"; on the other hand there is "a general asymmetry in knowledge: By themselves such foundations do not go far, but not knowing them can be a disaster (in a sense of this word suitable to the academic situation involved)."[23]

Kreisel defends formalisation with more specificity in his review [143] of Putnam's essay "Mathematics without foundations" [199]. In the essay Putnam claims the Completeness Theorem for first order logic, giving conceptually distinct equivalents of first order provability, provides evidence for the anti-foundationalism crafted in that essay.[24] Putnam invokes Σ_1^0-completeness to the same end, because Σ_1^0-completeness gives an equivalence between on the one hand number-theoretic truths of Σ_1^0 form, with validity of an effectively chosen Σ_1^0-statement on the other – two conceptually distinct characterisations of a fundamental concept. Kreisel takes Putnam to task for failing to mention that the latter equivalence does not extend to number-theoretic truths of Π_1 form, for which the equivalence in question certainly fails – Peano Arithmetic being Σ_1^0-complete but not Π_1^0-complete.

In the language of this book, Kreisel is chiding Putnam for overlooking the *entanglement* of number-theoretic truth with syntax, in the first order case; and for overlooking the entanglement of number-theoretic truth with a logic in the second order case. (Interestingly, Kreisel does not take Putnam to task for mentioning the completeness theorem in connection with anti-foundationalism.)

Or to put it another way, what Kreisel is noticing (for foundationalism) is just what Fer pointed out (for abstraction): the impossibility of keeping the twin poles of syntax and semantics apart for long.

---

[22] As is well known, Putnam went on to argue in the paper that the Löwenheim-Skolem theorem would then apply, undercutting the idea of univocal reference. In the time since the writing of that paper, a more conservative view of the prospect of formalising scientific theories in a first order way has emerged, as can be seen in the remark of Boris Zilber, that "a formal theorem like Löwenheim-Skolem could not undermine the quest for a univocal theory of physics". Zilber, lecture "Syntax, definability and geometry," Seminario de Lógica y Geometría, Bogotá, May 13, 2020, and personal communication.

[23] [140], p. 101.

[24] Of course the principal theme of [199] is the purported equivalence between the set-theoretical and the modal logical formalisations of mathematics.

4.2 Episodes of Formalism Independence
in Gödel's Writings

We saw that formalism independence was linked, for Gödel, to absoluteness. Absoluteness has a precise technical meaning. Viewed syntactically: given a theory T a formula in the language of T is *absolute* if and only if it is provably Δ_1 in the language of T, that is, provably equivalent to both a Σ_1 and a Π_1 formula in the given language.[25] If a sentence is absolute, if it is true in a model M of the theory it is true in all transitive extensions of M, being equivalent to an existential formula and therefore persisting upwards;[26] and conversely if it is true in a model M then it is true in all submodels of M, as it persists downwards, being equivalent to a universal formula.

Viewed semantically: given a formula ϕ in the language of set theory, we say that ϕ is *absolute* if given a transitive model M and elements $x_1, \ldots x_n$ of M,

$$\phi^M(x_1, \ldots x_n) \leftrightarrow \phi(x_1, \ldots x_n),$$

where in the right hand side of the equivalence it is asserted that ϕ holds in V, the cumulative hierarchy of sets.[27] What is interesting about the definition is that it equates two different notions of truth, to wit: the Tarskian notion of truth in a model and the notion, "true in V". The equivalence is opaque for one not equipped with or otherwise averse to the notion of truth in the cumulative hierarchy, there being no single truth definition serving for sentences of quantifier rank n for all n. On the other hand the stratified notion is clear: truth in V is definable for any quantifier rank n via a partial truth predicate for formulas of the given rank.[28] A consequence of the definition is that the absoluteness of ϕ is equivalent to the statement that for any transitive models $M \subseteq M'$ of ZFC, ϕ holds in M iff ϕ holds in M'.

Gödel used the term "absolute" in different ways. Sometimes he used it in the technical sense, and sometimes he applied the term more broadly, to indicate formalism independence *tout court*. Here is an excerpt from Gödel's abstract entitled "On the length of proofs":

---

[25] It is trivial that Δ_1 formulas are absolute. The converse is due to Feferman and Kreisel, see [59].

[26] This is the definition of the term "absolute" in set theory. In the context of arithmetic the term "absolute" would apply to Δ_1^0 concepts, presumably.

[27] See for example [113], p. 154.

[28] Alternatively truth in V_α, the α-th level of the cumulative hierarchy, is definable. In fact by the Levy Reflection Theorem, if ϕ holds then there is a club of ordinals α such that ϕ holds in V_α. Of course here the definition requires quantifying over a proper class, the class of ordinals.

Thus the notion 'computable' is in a certain sense 'absolute', while almost all metamathematical notions otherwise known (for example, provable, definable, and so on) quite essentially depend on the system adopted.[29]

The deep question Gödel's side remark raises in the abstract, which pre-occupied Post quite a bit as well, is whether the metamathematical notions 'provable' and 'definable' admit absolute formulations in the way that – in Gödel's view – 'computable' does.

Gödel would place the question at the centre of his 1946 Princeton Bicentennial Lecture. We treat the lecture at length below, after first turning briefly to some remarks from Gödel's 1929 thesis on the notion of "absolute provability".

Gödel's Notion of "Solvability by All Means Imaginable"

In his thesis Gödel distinguishes formal provability from what Gödel will in his later writings call absolute provability, or provability "through all specified means". More precisely, he observes that one might raise the following objection to the main result of his thesis, namely that the law of excluded middle in its proof may "invalidate the entire completeness proof". The Completeness Theorem asserts

"a kind of decidability," namely every quantificational formula is either provable or a counterexample to it can be given, whereas the principle of the excluded middle seems to express nothing other than the decidability of every problem.[30]

Thus the proof may be circular: one assumes the decidability of every question in order to prove just that assertion.[31]

But, Gödel goes on to say, what he has shown is the provability of a valid formula from "completely *specified, concretely enumerated* inference rules",[32] not merely from all rules imaginable; whereas the law of excluded middle is used informally in the sense that the notion of decidability or solvability asserted by the law is left unspecified. As Gödel puts it:

[29] [84], p. 399.
[30] [84], p. 63.
[31] This objection to the Law of Excluded Middle is the content of Brouwer's "Third Insight" as expressed in his "Intuitionistische Betrachtungen über den Formalismus". See p. 40, [166]. for an English translation.
[32] Gödel's emphasis.

... what is affirmed (by the law of excluded middle) is the solvability not at all
through specified means but only through all means that are *in any way imaginable*
... 33

The Completeness Theorem, then, provides a reduction: if we assume solvability by all means imaginable, then we have, in the case of a sentence of first-order predicate calculus, a reduction to solvability by very specific means laid out beforehand. Gödel remarks in a footnote to this passage that the notion of provability "by any means imaginable" is perhaps "too sweeping",[34] and in fact Gödel entertained the idea in the 1930s and early 1940s, that certain other mathematical problems may be absolutely *un*decidable. In 1939, for example, Gödel explained his consistency proofs of both the Axiom of Choice and the continuum hypothesis in a lecture in Göttingen, voicing his suspicion that the axiom of constructibility, $V = L$, would be absolutely undecidable:

The consistency of the proposition A (that every set is constructible [$V = L$]) is also of interest in its own right, especially because it is very plausible that with A one is dealing with an absolutely undecidable proposition, on which set theory bifurcates into two different systems, similar to Euclidean and non-Euclidean geometry.[35]

He will later discard the view. In a remark that is especially important for the interpretation of Gödel's general standpoint in this book, Gödel here takes the view that the Incompleteness Theorems do not, in the end, destroy the Hilbert programme "in its original extent and meaning":

As to problems with the answer Yes or No, the conviction that they are always decidable remains untouched by these results.[36]

That is to say, an informal decision procedure will eventually be developed and will yield decidability – restoring mathematics to its prelapsarian condition of being equipped with a bivalent notion of truth. Gödel would later expand on the idea of an informal decision procedure in a discussion about Leibniz's *characteristica universalis* with Carnap in 1948:

---

33 ibid.
34 See [84], p. 65.
35 [*1939b], in [87], p. 155.
36 [193?] in [85].

The universal characteristic claimed by Leibniz (1677) does not exist. Any systematic procedure for solving problems of all kinds would have to be nonmechanical.[37]

Kreisel entertained the idea of an informal decision procedure in 1972:

> [I]t has been clear since Gödel's discovery of the incompleteness of formal systems that we could not have mathematical evidence for the adequacy of any formal system; but this does not refute the possibility that some quite specific system . . . encompasses all possibilities of (correct) mathematical reasoning . . . In fact the possibility is to be considered that we have some kind of nonmathematical evidence for the adequacy of such [a system].[38]

Indeed one could argue that Kreisel began to develop such a procedure in the form of his squeezing arguments (see Section 5.3).

Returning to Gödel's Completeness Theorem, its main interest from the standpoint of this book, also (apparently) from the standpoint of Putnam's [199], is in showing that first order consequence can be characterised semantically, by revealing a correspondence between first order provability and a fundamentally semantic notion, validity.[39] Depending on their philosophical agenda, logicians could now restrict themselves wholly to syntax, or on the other hand wholly to semantics, without risk, as it were, of leaving out the other point of view.[40]

---

[37] Gödel Nachlass, folder 1/209, 013184, p.1. See [269] for the concept of "strong" absolute undecidability as well as for an extensive discussion of Gödel's views on absolute undecidability during the period of the 1930s and early 40s.

[38] Kreisel [138], p. 322. See also Kreisel [141].

[39] It is interesting to note that this way of viewing the theorem did not appear in print until Robinson's 1951 [207], an indication of the lack of focus on semantics characteristic of the logical work of the 1930s and 40s (see below).

[40] That was, at least, the idea – that a clear separation of syntax from semantics had been achieved. Thus E. Nelson could take the view that mathematics is "all syntax":

The mental image – semantics in its poetic role – behind the concept "x is infinitesimal" is the 17th century image of a number x so small that it is indeed the ghost of a departed quantity. But this mental image plays no ontological role; the existential statement "there exists an infinitesimal x" reduces to a commonplace about strictly positive numbers. The moral of this is that the role of syntax in mathematics is not to express semantic truths (because there are no semantic truths in mathematics to express). Mathematics is syntax, and the syntax of mathematics is mathematics itself. (E. Nelson, "Syntax and Semantics," www.math.princeton.edu/?nelson/papers.html)

Others, such as Poincaré, could argue for the ineliminability of semantic content; and Putnam could build the case for anti-foundationalism straightway, thus in some fashion avoiding the syntax/semantics distinction altogether.
Another virtue of the Completeness Theorem in this connection is that it shows that the concept of first order validity is absolute, due to the equivalence of the universally quantified or Π_1 statement saying that every model satisfies a given sentence, with a bounded or Δ_0

In spite of his having supplied the fundamental example, Gödel did not pursue the semantic reformulation of metamathematical concepts, at least not explicitly – though one could argue that the second, in Gödel's terms "more perspicuous" presentation of the constructible hierarchy in terms of closure of the ordinals under the so-called Gödel functions, was given in this spirit.[41] This is in contrast to Tarski's logical work, which was carried out as part of an explicit programme to provide semantic reformulations of metamathematical theorems, his 1929 characterisation of definable sets of real numbers being a prime example.

As for the idea of an informal decision procedure, Gödel lays out the beginnings of an informal proof system in his 1946 Princeton Bicentennial Lecture, also in his 1947 "What is Cantor's continuum problem?" [82]. The system is based on principles involving large cardinals, which are built over ZFC set theory, and which are *true*. We will return to this proof system in Section 4.3.1.

Dialectica Interpretation

The *"Dialectica Interpretation"*,[42] gives an interpretation of Heyting's intuitionistic arithmetic in terms of so-called computable functionals of finite type. This gives a consistency proof for Heyting arithmetic and thus for Peano Arithmetic, if one combines the *Dialectica* Interpretation with Gödel's earlier double-negation interpretation of Peano Arithmetic into Heyting Arithmetic.

In his introduction to [83],[43] Troelstra notes the *logic*-freeness of Gödel's system T:

> Gödel did not want to go as far as admitting Heyting's abstract notion of constructive proof; hence he tried to replace the notion of constructive proof by a more definite, less abstract (that is, more nearly finitistic) notion, his principal candidate being a notion of "computable function of finite type" which is to be accepted as sufficiently well understood to justify the axioms and rules of his system T, *an essentially logic-free theory* of functionals of finite type.[44]

In a crucial passage of the paper Gödel explains the move to replace the notion of a Heyting proof with a more concrete, perspicuous, but still abstract notion. The choice is forced by the Second Incompleteness Theorem, which

---

statement, with ω as parameter, asserting the existence of a natural number code of a proof. (Π_1 and Δ_0 in the Levy hierarchy.)

[41] See Section 4.3.2.
[42] [83].
[43] "Note to 1958 and 1972" in [85].
[44] [85], p. 221. Emphasis ours.

demonstrates that the consistency proof will require abstract notions going beyond what Gödel calls the "finitary attitude":

Here by abstract (or nonintuitive) we must understand those that are essentially of second or higher order, that is, notions which do not involve properties or relations of *concrete objects*, (for example, combinations of signs), but that relate to *mental constructs* (for example, proofs, meaningful statements and so on); and in the proofs we make use of insights, into these mental constructs, that spring not from the combinatorial (spatiotemporal) properties of the sign combinations representing the proofs, but only from their *meaning*.[45]

Relevant to our concerns here is that Gödel takes as primitive the notion of intuitionistic evidence, rather than the concept of a Heyting proof; for a consistency proof it is not enough to work with "(spatiotemporal) properties of the sign combinations representing the proofs"; one must work with the concept of meaning *directly*.

The presentation of the system T is obscure in places. As for avoiding the notion of constructive proof, in Gödel's later correspondence with Bernays, Bernays questioned whether the notion of constructive proof is not needed after all – is it not implicit in the concept of computable function of finite type?[46]

In an interesting (from our point of view) footnote to the paper Gödel draws a distinction between grasping content sufficiently clearly and recognising that axioms hold for it:

One may doubt whether we have a sufficiently clear idea of the content of this notion, but not that the axioms given below hold for it. The same apparently paradoxical situation also obtains for the notion, basic to intuitionistic logic, of a proof that is informally understood to be correct ... If the notion of computable function is not to implicitly contain the notion of proof, we must see to it that it is

---

[45] See Gödel's [1958], in [85], p. 241. Recall also the passage in Gödel's letter to Rappaport:
  Likewise it does not follow from my theorems that there are no convincing consistency proofs for the usual mathematical formalisms, notwithstanding that such proofs must use modes of reasoning not contained in those formalisms. What is practically certain is that there are, for the classical formalisms, no conclusive combinatorial consistency proofs (such as Hilbert expected to give), i.e. no consistency proofs that use only concepts referring to finite combinations of symbols and not referring to any infinite totality of such combinations. I have published lately (see Dial., vol. 12 (1958)a p. 280) a consistency proof for number theory which probably for many mathematicians is just as convincing as would be a combinatorial consistency proof, which however uses certain abstract concepts (in the sense explained in this paper). [89], pp. 176–177.

[46] See [88]. Another example of formalism freeness in the setting of constructive mathematics is Bishop's *Foundations of Constructive Analysis* [22], in which Bishop gives a relatively formalism free presentation of mathematical analysis in a constructive framework.

immediately apparent from the chain of definitions that the operations can be performed, as is the case for all functions in the system T specified below.[47]

Thus we see Gödel registering the point, that content somehow supersedes axiomatisation – a kind of "semantic vitalism", to stretch Gödel's terminology just a bit.

We now take up the main point of departure for the ideas in this book, Gödel's [86].

4.3 Gödel's Princeton Bicentennial Lecture

Gödel begins the lecture with the concept of computation, pointing out that this concept can be given a formalism independent definition:

> Tarski has stressed in his lecture the great importance (and I think justly) of the concept of general recursiveness (or Turing computability). It seems to me that this importance is largely due to the fact that with this concept one has succeeded in giving an absolute definition of an interesting epistemological notion, i.e. one not depending on the formalism chosen.[48]

Two notions of formalism independence are in play. One the one hand there is formalism independence in the sense of *confluence*, i.e. stability of a concept across conceptually distinct formal frameworks (as we saw in the previous chapter). For computability this is manifest in the fact that all of the known mathematical notions of computability that had surfaced since 1934, i.e. the Gödel–Herbrand–Kleene definition (1936), Church's λ-definable functions (1936), Gödel–Kleene μ-recursive functions (1936), Turing Machines (1936) and Post (1943) systems,[49] define the same class of functions, as we saw. But then a second sense of formalism independence emerges in the passage, having to do with *absoluteness* in the technical sense defined above, namely provably Δ_1^0 in a given theory. C. Parsons seems to interpret the paragraph in his introductory note to [86] in the former sense, in terms of "the absence of the sort of relativity to a given language that leads to stratification of the notion such

---

[47] [83], reprinted in [85], p. 243.
[48] [85], p. 150.
[49] see, e.g., [45].

as (in the case of definability in a formalized language) into definability in languages of greater and greater expressive power". The stratification is "driven by diagonal arguments".[50]

Subtleties here involving the absoluteness claim vis a vis the features of the various formalizations of the notion of computability have been pointed out by Sieg, who remarks in his [235] that Gödel's claim is problematically circular, resting as it does on an absolute notion of computable function. This is because the notion of formal system is to be defined in terms of the notion of absolute computability, whereas the notion of formal system seems to be needed in order to give the proof of absoluteness. In the previous chapter we argued at length that this circularity is obviated by the fact that Turing's model provides a *grounding*.

In a footnote appended to this first sentence in 1965,[51] Gödel offers the following clarification of it:

> To be more precise, a function of integers is computable in *any* formal system containing arithmetic if and only if it is computable in arithmetic, where a function f is called computable in S if there is in S a computable term representing f.[52]

Gödel is referring to the fact that any computable (i.e. partial recursive) function representable by a term in a formalism *extending* arithmetic, is already representable in Peano arithmetic, by the same term. The proof depends on Kleene Normal Form (defined in Section 3.3), and generalises the remark Gödel made in the abstract for his Speed-up Theorem cited at the end of Section 3.2, about the absoluteness of computability relative to higher order extensions. The 1965 clarification was needed, as Gödel seems to have conflated two different concepts. "Formalism independence" taken in the sense of "confluence" applies to a broad range of conceptually and/or formally distinct notions of computability, and is established by proving the equivalence of the corresponding encodings of these various notions in a suitably chosen metatheory.[53] Whereas in the technical sense given (possibly) in the lecture and subsequently in the 1965 footnote, absoluteness is a property of a concept relative to a particular class of formal systems. Strictly speaking, absoluteness in this sense is a restricted form of confluence, applying "piecewise", so to speak, to particular notions of computability, which are only absolute relative to (a particular class of) their extensions.

---

[50] Parsons, introductory note to Gödel's [1946], in [85], p. 145.

[51] to the version of the lecture published in [47].

[52] See also Gödel's addendum to *1936a* cited in Section 2.1. Emphasis ours.

[53] A rather weak metatheory is required for this.

In any case the observation here is that the concept of partial recursive function is in some sense saturated for arithmetic; that is to say, there is no computable (i.e. partial recursive) function which is computable in a system *extending* arithmetic, which is not already computable in arithmetic. To be computable in one arithmetic formal system is to be computable in all of them.[54]

Gödel contrasts this situation with the apparently less felicitous cases of provability and definability:

> In all other cases treated previously, such as demonstrability or definability, one has been able only to define them relative to a given language, and for each individual language it is clear that the one thus obtained is not the one looked for. For the concept of computability, however, although it is merely a special kind of demonstrability or definability, the situation is different. By a kind of miracle it is not necessary to distinguish orders, and the diagonal procedure does not lead outside the defined notion.

We see here Gödel revisiting a comment in his 1936 abstract "On the length of proofs", that the (known) notions of provability and definability are language-relative. He also revisits the idea of computability as a form of deduction – the lesson, for him, of the Turing Machine – declaring computability to be also a form of definability, which indeed it is, though it is not always thought of this way. Gödel then makes the bold move of asking for an analysis of the two remaining concepts (demonstrability and definability) along the lines of the analysis that Turing gave for computability:

> This, I think, should encourage one to expect the same thing to be possible also in other cases (such as demonstrability or definability). It is true that for these other cases there exist certain negative results, such as the incompleteness of every formalism ... But close examination shows that these results do not make a definition of the absolute notions concerned impossible under all circumstances, but only exclude certain ways of defining them, or at least, that certain very closely related concepts may be definable in an absolute sense.[55]

Gödel goes on to say that just as for computability, where there is an ambient intuitive (epistemological) notion to be made precise – human calculability by a fixed routine – the intuitive notion to be made precise in the case of definability is "comprehensibility by the mind". For provability, one assumes that

---

[54] Note that the terminology here stops short of the language of transcendence Gödel would use later.

[55] [84], p. 399.

the intuitive (epistemological) concept Gödel had in mind was some notion of informal consequence.[56]

It is very striking that Post had the same thought, albeit somewhat later, of grouping together the three notions, computability, definability and provability, and then asking for an absolute formulation of each.[57] The thought is a natural one: each of these notions come with their own paradoxes that, with some care, can be turned into theorems. Thus it makes sense to ask for absolute or at least paradox-immune notions of each. We will see that Gödel proposed hereditary ordinal definability as a possible absolute notion of definability in his 1946 lecture. Post arrived at the same concept independently between 9 September 1952 and 4 February 1953, the date of a Sterling notebook in which he recorded reflections on the notion of definability.[58]

We now briefly consider Gödel's suggestions regarding provability, before turning to definability.

4.3.1 Provability

Let us consider, e.g. the concept of demonstrability. It is well known that, in whichever way you make it precise by means of a formalism, the contemplation of this very formalism gives rise to new axioms which are exactly as evident and justified as those with which you started, and that this process of extension can be extended into the transfinite. *So there cannot exist any formalism which would embrace all these steps*; but this does not exclude that all these steps... could be described and collected in some non-constructive way. In set theory, e.g., the successive extensions can be most conveniently represented by stronger and stronger axioms of infinity. It is certainly impossible to give a combinational and decidable characterization of what an axiom of infinity is; but there might exist, e.g., a characterization of the following sort: An axiom of infinity is a proposition which has a certain (decidable) formal structure and which in addition is true. Such a concept of demonstrability might have the required closure property, i.e., the following could be true: Any proof for a set-theoretic axiom in the next higher system above set theory (i.e. any proof involving the concept of truth which I just used) is replaceable by a proof from such an axiom of infinity. It is not impossible that for such a concept of demonstrability some completeness theorem would hold which would say that every proposition expressible in set theory is decidable from the present [ZFC] axioms plus some true assertion about the largeness of the universe of all sets.[59]

---

[56] As for the term "epistemological", Gödel applied this term very broadly, often interchangeably with the term "philosophical". See for example the different uses of the term in Wang's [280].

[57] See Post's Collected Works [197]. See also A. Urquhart's [258], p. 469.

[58] Post seems to have been wholly unaware of Gödel's lecture. Post Nachlass, Philadelphia archives of the American Philosophical Society.

[59] [86] p. 151. Emphasis ours.

Axioms of infinity are thus a means of transcending the relevant hierarchies of formal languages. More concretely, some suitable hierarchy of large cardinal assumptions should replace the hierarchy of formal systems generated by, e.g., the addition of consistency statements to set theory, i.e. passing from ZFC to ZFC + Con(ZFC) and then iterating this; or the addition of a satisfaction predicate for the language of set theory, then considering set theory in the extended language, and iterating this.

The usefulness of transfinite concepts was mentioned by Gödel in a number of other contexts, e.g. that they complete the partial proofs of the Completeness Theorem given by Löwenheim and Skolem.[60] The claim is made here in an embryonic form of what came to be known as Gödel's programme for large cardinals, a programme laid out in its fullest form in his 1947 "What is Cantor's continuum problem?" [82].[61]

Returning to the above quote, it is indeed impossible to give a general definition of the concept of large cardinal, even so that the large cardinal axioms known nowadays can be stated in a first order language one by one. The proof system Gödel proposes here, consisting of ZFC together with axioms of infinity, or alternatively axioms about the largeness of the universe of sets, is then necessarily informal, tethered as it is to the (informal) concept of truth. H. Woodin expressed a similar thought in connection with his continuation of Gödel's programme for large cardinals, remarking that "truth [is] beyond our formal reach"; and that

> There is a component in the evolution of our understanding of mathematics which is not formal.
> There is mathematical knowledge which is not based on proofs.[62]

In detail: Gödel's remark that "It is certainly impossible to give a combinational and decidable characterisation of what an axiom of infinity is" follows from the Incompleteness Theorem, for otherwise if there were such a characterisation it could be used to show that ZFC can prove its own consistency, using inaccessible cardinals. There is also the more basic reason that truth is not definable in such systems, so a characterisation of a *true* axiom of infinity cannot be given. As for decidability, note that this can be delivered (trivially) by appending a large cardinal axiom of the form, e.g., $\kappa \neq \kappa$ to ZFC. What is wanted of course is a *true* large cardinal axiom. Of course one can prove the

---

60 See Gödel/Wang correspondence in [89], e.g. the letter of 7 March 1964 on p. 404.

61 In fact very few large cardinals had been discovered by 1946.

62 H. Woodin, *The Joint Quest for Absolute Infinity and the Continuum – From Cantor to Woodin*, University of Turku 2019.

consistency of large cardinals, assuming their existence in the first place. But now we have travelled some distance from what Gödel was asking for, that is we have gone from truth to consistency.

As for formalism independence, a partial attempt to replace logical hierarchies by infinitary principles, and thereby gain decidability, can be seen in the following result of H. Woodin: in the presence of large cardinals,[63] the Σ_1^2 theory of real numbers, i.e. existential statements about sets of reals, is "generically absolute" or (set) forcing immune, in the sense that their truth cannot be changed by forcing – one important form of decidability.[64]

In fact generic absoluteness has emerged as a beautiful form of invariance in set theory. Prior to the advent of forcing Shoenfield proved his absoluteness lemma [232] stating that any Π_2^1 statement that is true in L is true in every transitive model of set theory. With the advent of forcing it immediately became clear that arithmetic truths cannot be changed by forcing, because the natural numbers are absolute. But forcing clearly introduces a great deal of variability otherwise. What generic absoluteness does is to disable the forcing mechanism, eliminating set-theoretic variability of this kind.

It is an open problem, whether under some large cardinal assumption and $\diamondsuit$ (or something like it) any Σ_2^2-statement that can be made true by forcing, is actually true. With general set-theoretical statements, early researchers experimented with principles of the form: any statement that can be forced to be true is actually true. This is of course inconsistent as one can always force the continuum hypothesis (henceforth CH) to be true, or on the other hand one can always force $\neg CH$ to be true. An early attempt in the direction of a modified principle of this kind is the principle MAX(CCC) ([243]) which says that any statement with parameters of hereditary cardinality less than the continuum, which can be made true by CCC-forcing,[65] and is provably preserved by CCC-forcing, is already true.

Is the proof concept associated to a generically absolute theory a candidate for an absolute notion of provability? An Ω-proof, a proof concept which occurs in Woodin's work in connection with generic absoluteness,[66] is just a universally Baire set of reals. Replacing the concept of a proof by a universally Baire set of reals has an appearance of formalism freeness, somewhat

---

[63] A proper class of Woodin cardinals.

[64] See [288]. The result requires the Continuum Hypothesis. Another result of this kind due to Woodin says about the structure $L(\mathbb{R})$, the constructible closure of the reals, that its first order theory is (set) forcing absolute in the presence of large cardinals (a proper class of measurable Woodin cardinals), [287].

[65] CCC comes from "Countable Chain Condition" and means that the partial order used in the forcing does not contain uncountable antichains.

[66] [288].

reminiscent of the idea of replacing a formula by a set invariant under automorphisms in the abstract elementary class context. Universally Baire sets of
reals are also very *robust*. Like the concept of recursive function, they can be
defined in many conceptually different ways.[67]

There is also the important question whether forcing is completely general.
Or to put it another way, how do we know that another model construction
technique will not materialise, introducing new forms of variability? Matteo
Viale presented in the Vienna ESTC in 2019 the following unpublished result,
a step toward proving that forcing is a completely general method:

Theorem 4.3.1 *(Viale) For suitable initial fragments M of the universe of sets
V any elementary superstructure of M is a substructure of a model of the form
V^B / G for some complete Boolean algebra $B \in V$ and ultrafilter G on B.*

In the previous chapter we isolated a two part analysis involving the *confluence* of distinct formalisations of a concept, together with a grounding
example. The notion "universally Baire set of reals", for example, has many
distinct characterisations, as was noted. In fact confluence in set theory is ubiquitous, the most notorious recent case involving the fact that many conceptually
distinct principles imply that the value of the continuum is $\aleph_2$. Here are some
of them:

- Woodin's $(\star)$-axiom says that the Axiom of Determinacy holds in
  $L(\mathbb{R})$ and $L(P(\omega_1))$ is a $\mathbb{P}_{max}$-generic extension of $L(\mathbb{R})$ [288].
  Implies continuum is $\aleph_2$.
- Martin's Maximum (MM) says that if P is a forcing notion which preserves
  stationarity of subsets of ω_1 and D is a set of open dense subsets of P,
  $|D| \leq \aleph_1$, then there is a filter subset of P which meets every dense set in
  D [70]. Implies continuum is $\aleph_2$.
- Mapping Reflection Principle (MRP) [179]. Implies continuum is $\aleph_2$.
- Bounded Proper Forcing Axiom (BPFA) i.e. if P is a forcing notion which
  preserves stationarity of subsets of $[\lambda]^\omega$ for all regular uncountable λ, and D
  is a set of open dense subsets of P of cardinality $\leq \aleph_1$ such that $|D| \leq \aleph_1$,
  then there is a filter subset of P which meets every dense set in D [91].
  Implies continuum is $\aleph_2$.
- There is an extension of Lebesgue measure to a countably additive measure
  defined on every set of reals [257]. Implies continuum is greater than $\aleph_1$, in
  fact at least as great as the first weakly inaccessible cardinal.

---

[67] See Feng, Magidor and Woodin's [63].

- Stationary Basis Hypothesis: There exists ω_1 many stationary sets, $\langle S_\alpha : \alpha < \omega_1 \rangle$, such that for every stationary set $S \subseteq \omega_1$, there exists $\alpha < \omega_1$ such that $S_\alpha \subseteq S$ modulo the nonstationary set [225]. Implies continuum is at least $\aleph_2$.

Some large cardinal properties exhibit a great deal of confluence – a concept which can serve as a marker of formalism freeness. The following are equivalent for strongly inaccessible κ:

- κ is weakly compact, i.e. the infinitary language $L_{\kappa\omega}$ has the κ-compactness property: if a theory of cardinality $\leq \kappa$ has the property that every subset of size $< \kappa$ has a model, then the theory itself has a model.
- $\kappa \rightarrow (\kappa)^2_2$, i.e. if 2-element subsets of κ are coloured by two colours, red and blue, than there is a subset H of κ of cardinality κ such that every pair from H is red or every pair from H is blue.
- Every κ-tree (a tree with height κ where every level is of cardinality $< \kappa$) has a branch of length κ.
- κ is Π^1_1-indescribable, i.e. if $A \subseteq \kappa$ and ϕ is a Π^1_1-sentence true in $(V_\kappa, \in, A)$, then there is $\lambda < \kappa$ such that ϕ is true in $(V_\lambda, \in, A \cap \lambda)$.
- For all $A \subseteq V_\kappa$ there are a transitive set M and $B \subseteq M$ such that $\kappa \in M$ and $(V_\kappa, \in A) \prec (M, \in, B)$.

The following are equivalent for strongly inaccessible κ:

- κ is measurable (i.e. there is a 2-valued κ-additive non-trivial measure on $\mathcal{P}(\kappa)$).
- κ is real-valued measurable (i.e. there is a real-valued κ-additive non-trivial measure on $\mathcal{P}(\kappa)$).
- There is a κ-complete non-principal ultrafilter on κ.
- There is an elementary embedding $i : V \rightarrow M$, where M is a transitive class, $i(\alpha) = \alpha$ for $\alpha < \kappa$, and $i(\kappa) > \kappa$.

4.3.2 Definability

Gödel now passes to definability, in connection with which he can give us "somewhat more definite suggestions".

The idea, again, is to obtain an absolute notion of definability. Such a goal seems paradoxical, as definability appears to be rather closely tied to a formalism, in the sense that one uses a signature and formation rules to build up well-formed formulas, and then the definable sets are simply declared to be those sets that are definable via that stock of formulas. But the standard technical definition of definability is not what Gödel has in mind. For Gödel, definability is an epistemological notion. Finding an absolute characterisation of it is to find "an adequate formulation for comprehensibility by our mind".[68]

The link between definition and comprehensibility (or even knowledge) can be traced back to the classical period[69] – indeed even the word *logos* can be translated as *definition*.[70] The classicist Rose Cherubin traces the link to Aristotle:

> [In Aristotle's] Metaphysics Book I Chapter 3, 983a24ff., in the discussion of why those seeking wisdom need to acquire knowledge of first causes and principles, he [Aristotle JK] notes that we say that we understand each thing when we think that we know its first (i.e. most fundamental) cause; and that causes "are said in four ways." 'Cause' translates 'aitia' which means 'that which is responsible for something' or 'the why of something.' 'Said in four ways' seems to mean that when we speak of causes, we refer to one or more of four kinds of thing.
>
> One of the kinds of thing we invoke or refer to by means of the word 'cause,' Aristotle says, is the 'ousia' (substance, being) of a thing, or 'to ti en einai,' what it is to be that thing (sometimes translated as 'essence'). This is because "that through which" [the thing is as it is] "leads back to the ultimate formula"; 'formula' = 'logos.'
>
> This is why this kind of cause came to be called 'formal cause' and why it came to be associated with definition.
>
> ... [In Posterior Analytics 71b10ff. and 94a20ff.] Aristotle says that we think that we know a thing when we know (or think we know) its cause; the latter example provides an illustration of knowing that comes through knowing the logos-type cause.[71]

Gödel is forging a similarly robust connection between definition and knowledge by classifying definability under the heading of *epistemology*. The epistemological framework is needed: simply laying down a language and then declaring that the sets "comprehensible by our mind" to be those given by the formulas of the language, not only introduces arbitrariness; it leads, as Gödel will later observe,[72] to paradox.

After these initial remarks Gödel introduces the concept of "ordinal definability" and implicitly the concept of hereditary ordinal definability, denoted HOD. The idea is to take the ordinals as already given. A set is then ordinal definable if it is definable by a formula in the language of set theory with

---

[68] [85], p. 152.

[69] I am grateful to A. Arana for pointing out the connection between knowledge and definition in Aristotle in his 2019 CLMPS lecture, Prague and in private correspondence.

[70] The word *horismos* can also be so translated.

[71] R. Cherubin, personal communication.

[72] Observe to Wang, that is, citing the history of Frege's Axiom V. See [280].

finitely many ordinal parameters; it is hereditarily ordinal definable if it is ordinal definable, all its elements are ordinal definable, and so on.[73]

> Here you also have, corresponding to the transfinite hierarchy of formal systems, a transfinite hierarchy of concepts of definability. Again it is not possible to collect together all these languages in one, as long as you have a finitistic concept of language, i.e as long as you require that a language must have a finite number of primitive terms. But, if you drop this condition, it does become possible ... by means of a language which has as many primitive terms as you wish to consider steps in this hierarchy of languages, i.e. as many as there are ordinal numbers. The simplest way of doing it is to take the ordinals themselves as primitive terms. So one is led to the concept of definability in terms of ordinals ... This concept should, I think be investigated.

The advantage of doing things this way is that ordinals bequeath their "lawlikeness" to the sets constructed from them, namely the ordinal definable sets. They are "formed according to a law".[74]

Definability simpliciter is not definable in set theory, as we have noted. But ordinal definability is definable in set theory – something Gödel must have known, judging from his remarks here. Modern proofs of this depend on the Levy Reflection Principle, which was only proved in 1960. For this reason, if one passes to the "next language", i.e. one obtained by adding a truth predicate for statements about ordinal definable sets, one obtains no new ordinal definable sets.[75] Ordinal definability, in analogy with the computable functions, is "non-diagonalisable".

In what sense is ordinal definability formalism independent? In fact similar to constructibility, by a result of [276] the class of hereditarily ordinal definable sets can be obtained as the closure of all sets of the form V_α under the Gödel operations, instead of via the first order language.[76] This is an apparently formalism free construction, relative to set theory.[77]

Gödel's goal in these remarks involves replacing a formalism, or more precisely a hierarchy of them generated by the addition of truth predicates, by an axiom or principle of definability of a special kind: the characterisation of the principle must be decidable, and the principle must be *true*. The principle

---

[73] Technically: a set is hereditarily ordinal definable if all the elements of its transitive closure are ordinal definable.

[74] [85], p. 152.

[75] More exactly, for any formula $\phi(x)$ there is another formula $\phi'(x)$ which says that $\phi(x)$ is true in HOD. So no new ordinal definable sets are created by referring to truth in HOD.

[76] [187]. Here $V_0 = \emptyset$, $V_{\alpha+1} = \mathcal{P}(V_\alpha)$ and $V_\nu = \cup_{\alpha<\nu} V_\alpha$ for limit ν.

[77] HOD is not absolute however (consistently $\text{HOD}^{\text{HOD}} \neq \text{HOD}$); HOD is "forcing fragile" in the sense that HOD can be different in different models obtained by forcing.

which is implicit in the concept of ordinal definability[78] – the Levy Reflection Principle[79] – would satisfy both requirements for Gödel. We noted that Gödel does not state this principle in the lecture, which is essential for showing that HOD is a definable class. The Levy Reflection Principle is not itself an axiom of *infinity* per se, in fact it is provable; but if it is slightly strengthened and then reflected to some V_α, it becomes an axiom of infinity.[80]

Gödel next takes up the idea of capturing a different notion of definability in a formalism independent, lawlike and definable way, namely through *constructibility*. The presentation of the constructible hierarchy in terms of definability was given by Gödel in 1939 in his original monograph on the consistency of the continuum hypothesis [80].

In 1940 he gave a second presentation of the constructible sets, as the closure of the class of ordinals under the so-called "Gödel operations" [81], as we noted in Chapter 1:

$$L_0 = \emptyset$$
$$L_{\alpha+1} = \{\mathcal{F}_n(X, Y) : X, Y \in L_\alpha \cup \{L_\alpha\}, 1 \le n \le 8\}$$
$$L_\nu = \bigcup_{\alpha < \nu} L_\alpha.$$

This latter, in Gödel's words more perspicuous presentation, with no satisfaction or definability predicates occurring in it, *is completely formalism free, being given in natural language, with no distinction drawn between syntax and semantics* (using the possibly more restricted characterisation of formalism freeness).[81]

Gödel does not see L as exemplifying an, in his sense, absolute notion of definability in any case, even so that, as in the case of ordinal definability, the constructible hierarchy is non-diagonalisable in the following sense: if we form the constructible hierarchy and then add to the language of set theory a predicate for "x is constructible", we do not obtain any new constructible sets. Gödel continues:

> but, comparing constructibility with the concept of ordinal definability just outlined, you will find that not all logical means of definition are admitted in the definition of constructible sets. . . . *This has the consequence that you can actually define sets, and even sets of integers, for which you cannot prove that they are constructible* (although this can of course be consistently assumed). For this

---

[78] actually HOD here.

[79] Recall that the principle says that for every n there are arbitrarily large ordinals α such that $V_\alpha \prec_n V$.

[80] There is an α such that for all $A \subseteq \alpha$ there is $\beta < \alpha$ with $(V_\beta, \in, A \cap V_\beta) \prec (V_\alpha, \in, A)$. This implies that α is (strongly) inaccessible [116, Proposition 6.2].

[81] As was mentioned above, the L-hierarchy is then obtained by closing the ordinals under the so-called Gödel functions. See Section 2.2.1.

reason, I think constructibility cannot be considered a satisfactory formulation of definability.[82]

The remark is astonishingly prescient. How will we find that not all sets are constructible? This is shown by Cohen forcing, which emerged only two decades later.

We will return to constructibility below. It turns out that the constructible hierarchy is very robust and permits interesting generalisations. As for HOD, Gödel predicts the consistency, proved later by McAloon [174], of the axiom $V = \text{HOD} + 2^{\aleph_0} > \aleph_1$.

At the end of the address Gödel remarks of his two candidates for the concept of absolute definability – constructibility and ordinal definability – that neither of these is an absolute notion in the sense of the paper:

> ... in both examples I gave, [ordinal definability and constructibility] the concepts arrived at or envisaged were not absolute in the strictest sense, but only with respect to a certain system of things, namely the sets as conceived in axiomatic set theory; i.e., although there exist proofs and definitions not falling under these concepts, these definitions and proofs give, or are to give, nothing new within the domain of sets and propositions expressible in terms of "set," "∈," and the logical constants.[83]

It would seem that, for Gödel, an absolute notion of definability "in the strictest sense" would have to transcend the background theory – axiomatic set theory, in this case. This is slightly paradoxical in that on the one hand, Gödel's use of the word "absolute" seems to indicate a desire for a characterisation of definability which is not tied to the background theory; while on the other hand he is clearly reluctant to attach a fully transcendental concept, i.e. one not definable in set theory, to the "epistemological" notion in question. So while aspiring to absoluteness in the "strictest sense",[84] Gödel is, at the end of the day, committed to set theory as his metatheory. Idealisation is decisive here, Gödel will say later,[85] in connection with the constructible hierarchy. But just as with constructibility, idealisation does not have to bring full transcendence.

That this is the fundamental point of tension here bears repeating: Gödel aspires toward absoluteness in the strictest sense, that is to say we read Gödel as taking naive set theory as his metatheory, that is set theory informally construed. While on the other hand the measure of absoluteness in the technical

---

[82] Emphasis ours. [85], p. 152.
[83] ibid.
[84] as does Tarski, as we noted above.
[85] to Wang, [280], remark 8.3.3.

sense – our compass, so to speak – involves, in this case at least, the ZFC axiomatisation.

In Section 4.4 we propose an implementation of the ideas about definability in Gödel's 1946 Princeton Lecture, along the lines of extended constructibility.

4.3.3 Kantian Ideas

In *Logical Journey* [280], Hao Wang records the following exchange with Gödel on the topic of absolute definability and absolute provability:

> Once I asked Gödel about his Princeton lecture of 1946, in which he had discussed the task of extending the success of defining the concept of computability independently of any given language to "other cases (such as demonstrability and definability)." He replied:
> "Absolute demonstrability and definability are not concepts but inexhaustible [Kantian] ideas. We can never describe an Idea in words exhaustively or completely clearly. But we also perceive it, more and more clearly. This process may be uniquely determined – ruling out branchings."[86]

How did Gödel understand the notion, "Kantian idea"?

> The concept of *concept* and the concept of *absolute proof* [briefly, AP] may be mutually definable. What is evident about AP leads to contradictions which are not much different from Russell's paradox. Intuitionism is inconsistent if one adds AP to it. AP may be an *idea* [in the Kantian sense]: but as soon as one can state and prove things in a systematic way, we no longer have an idea [but have then a concept]. It is not satisfactory to concede [before further investigation] that AP or the general concept of concept is an idea. The paradoxes involving AP are intensional – not semantic – paradoxes. I have discussed AP in my Princeton bicentennial lecture.[87]

What distinguishes concepts from ideas? Earlier Wang records this exchange:

> Sometimes Godel hinted at a distinction between *concepts* and *ideas* along Kantian lines. On different occasions he spoke of the concepts of *concept, absolute proof*, and *absolute definability* as ideas rather than concepts:
> "The general concept of *concept* is an *Idea* [in the Kantian sense]. The intensional paradoxes are related to questions about Ideas. Ideas are more fundamental than concepts." (8.4.20)

---

[86] [280], remark 8.4.21, p. 268.
[87] 6.1.13, [280], p. 188.

There appears to be some disagreement among Kant scholars about what Kantian ideas are. Following M. Rohlf's account of them [210], they "represent *imaginary* focal points (A 644/B 672) that guide our study of nature and help us to achieve a more extensive and interconnected system of scientific knowledge".[88] This is because they are ideas of reason, and as such they are directed toward transcendental objects such as the soul, or "world-whole" – objects, in other words, beyond experience and therefore beyond the realm of what we can have knowledge of. As Rohlf writes:

> Moreover, some of these ideas – the transcendental ideas of the soul, the world-whole, and God – inevitably produce the illusion that we have *a priori* knowledge about objects corresponding to them. This putative knowledge seems to satisfy reason's demand for the unconditioned, though in fact these "**are only ideas**" that do not give us knowledge about transcendent objects but only produce the illusion of doing so (A 329/B 385).[89]

The Platonism usually attributed to Gödel would abjure the idea of reason having a built-in structural capacity for error. It is not surprising, then, that Gödel acknowledges the conflict – "It is not satisfactory to concede [before further investigation] that AP or the general concept of concept is an idea."

For Kreisel absolute definability and absolute provability are incalcitrant notions that "preoccupied (or paralyzed) Gödel since his lectures at Princeton".[90] But perhaps Gödel can be forgiven for his preoccupations. As it turns out, the concept of hereditary ordinal definability is at the very centre of current research in set theory. And while forcing could well have paralysed the *set theorist's* search for an intended model, the natural language paradigm Gödel laid down in his work of 1946–7, in which one draws on the concept of "true large cardinal axiom" in the pursuit of decidability in set theory, showed that there was a way forward.

4.4 Implementation

Inspired by Gödel's call for finding a formalism independent concept of definability, we offer a possible implementation specialised to the context of set theory. The move made here is an instance of a broader schema, based on a

---

[88] [210], p. 214. Emphasis ours.
[89] ibid, p. 200.
[90] [139], p. 4.

parametric use of logics, and it is as follows: consider a canonical mathematical object or construction. From the foundational point of view the object is built over a logic, generally first or second order logic. Now vary the underlying logic. Does the object change, or is the object invariant under the permutation of the underlying logic? We introduced a specific terminology in Chapter 1: we say that changes in the formal environment caused by small changes in, e.g. syntax, are evidence of *entanglement*; whereas a structure's indifference to changes of the underlying logic indicates formalism independence or *formalism freeness*.

One of the aims of this book is to present calculi for measuring degrees of entanglement/formalism freeness. We now present a calculus that builds on Gödel's two notions of definability, namely constructibility, and HOD. The calculus is a result of joint work with M. Magidor and J. Väänänen, and was introduced in [128]. We now present the methodology and some of the main results.

Fix a notion of definability, in this case constructibility, and view this notion of definability as an operator on a class of logics. We denote the result of applying this operator to a logic $\mathcal{L}^*$ by $L(\mathcal{L}^*)$, defined as follows:

$$
\begin{aligned}
L_{\alpha+1}(\mathcal{L}^*) &= \{X \subseteq L_\alpha(\mathcal{L}^*) : X \text{ is } \mathcal{L}^*\text{-definable} \\
&\quad \text{over } (L_\alpha(\mathcal{L}^*), \in) \text{ with parameters.}\} \\
L_\nu(\mathcal{L}^*) &= \bigcup_{\alpha < \nu} L_\alpha(\mathcal{L}^*) \text{ for limit } \nu \\
L(\mathcal{L}^*) &= \bigcup_\alpha L_\alpha(\mathcal{L}^*).
\end{aligned}
\tag{4.1}
$$

If $\mathcal{L}^*$ is taken to be first order logic, denoted $\mathcal{FO}$, we obtain the constructible hierarchy itself:

$$
\begin{aligned}
L_{\alpha+1} &= \{X \subseteq L_\alpha : X \text{ is first order definable} \\
&\quad \text{over } (L_\alpha, \in) \text{ with parameters.}\} \\
L_\nu &= \bigcup_{\alpha < \nu} L_\alpha \text{ for limit } \nu \\
L &= \bigcup_\alpha L_\alpha.
\end{aligned}
\tag{4.2}
$$

Myhill and Scott showed in [187] that if $\mathcal{L}^*$ is taken to be second order logic, denoted $\mathcal{SO}$, the class obtained is HOD, the hereditarily ordinal definable sets.[91]

In another precedent, the so-called Chang model $L(\mathcal{L}_{\omega_1\omega_1})$ is a model of ZF together with the failure of the Axiom of Choice, under large cardinal

---

[91] In fact this result enjoys some robustness, i.e. ostensibly much weaker logics than second order logic still give rise to HOD. See [128].

assumptions. In the light of this failure one might advocate the use of fragments of $\mathcal{SO}$, also on the basis of these having a somewhat more reasonable syntax.

It is not difficult to see that if $\mathcal{L}^*$ is taken to be weak second order logic $\mathcal{L}_w^2$, i.e. the logic allowing quantification over finite sets, we again obtain the constructible hierarchy L, that is, $L(\mathcal{FO}) = L(\mathcal{L}_w^2) = L$. This is interesting, since from the Lindström characterisation point of view these two logics are very different. In particular, $\mathcal{L}_w^2$ is non-compact.

Continuing with our observations about L-invariance, it is also not difficult to see that if one takes $\mathcal{L}^*$ to be that obtained from first order logic by adding to it the quantifiers Q_α, i.e. "there are at least $\aleph_\alpha$ many", for all cardinals $\aleph_\alpha$, one again obtains L.[92] Our eventual focus is on fragments of second order logic, and in fact when only cardinals that are second order characterisable are used, this logic will indeed be a fragment of second order logic.[93]

The following *implementation* of Gödel's suggestion in the 1946 lecture emerges. Define the equivalence relation on logics:

$$\mathcal{L}^* \equiv \mathcal{L}^{**} \text{ if and only if } L(\mathcal{L}^*) = L(\mathcal{L}^{**}).$$

This equivalence relation partitions the family of all logics into classes inside which the constructible hierarchy L is indifferent to what logic is used. Conceivably the equivalence classes could be small, which we would then interpret by saying that L seems to be quite dependent on the formalism used. As it happens, the classes are big, the class of $\mathcal{FO}$ including at least $\mathcal{L}_w^2$, i.e. the logic allowing quantification over finite sets, the family of extensions of $\mathcal{FO}$ given by $\mathcal{L}(Q_\alpha)$ for each $\aleph_\alpha$, and finally Magidor–Malitz logic, assuming $0^\sharp$ exists.[94] L in that sense "reads" all of these logics, including first order logic with the Magidor–Malitz quantifier adjoined to it, as first order. This is counterintuitive, given the fact that a number of these logics are far from being first order from the point of view of the Lindström characterisation.

---

92 See [128].

93 A cardinal κ is second order characterisable if there is a second order sentence ϕ of the empty vocabulary such that for all models M of the empty vocabulary, M has cardinality κ iff $M \models \phi$. All the $\aleph_n's$ are second order characterisable. If κ is second order characterisable, then so are $\kappa^+ 2^\kappa, 2^{2^\kappa}$, etc. See [266].

94 The Magidor–Malitz quantifier $Q_\alpha^{MM,n} x_1, \ldots, x_n \phi(x_1, \ldots, x_n)$ is defined as follows:
$\mathcal{M} \models Q_\alpha^{MM,n} x_1, \ldots, x_n \phi(x_1, \ldots, x_n) \iff \exists X \subseteq M (|X| \geq \aleph_\alpha \wedge \forall a_1, \ldots, a_n \in X :$
$\mathcal{M} \models \phi(a_1, \ldots, a_n))$. It turns out that if $\mathcal{L}^*$ is taken to be first order logic with the Magidor–Malitz quantifiers adjoined to it, then also $\mathcal{L}^* \equiv \mathcal{FO}$, assuming $0^\sharp$ exists. This is count erintuitive, since Magidor–Malitz logic is very far from first order logic as measured by the Lindström characterization. See [128] for details.

In fact, more is true:[95] The constructible hierarchy L is unaffected if first order logic is enriched in the construction of L by *any* of the following logical operations, simultaneously or separately:

- Recursive infinite conjunctions $\bigwedge_{n=0}^{\infty} \phi_n$ and disjunctions $\bigvee_{n=0}^{\infty} \phi_n$.
- Cardinality quantifiers $Q_\alpha, \alpha \in On$.
- Equivalence quantifiers[96] $Q_\alpha^E, \alpha \in On$.
- Well-ordering quantifier $\mathcal{M} \models Wx, y\phi(x, y) \iff \{(a, b) \in M^2 : \mathcal{M} \models \phi(a, b)\}$ is a well-ordering.
- Recursive game quantifiers $\forall x_0 \exists y_0 \forall x_1 \exists y_1 \ldots \bigwedge_{n=0}^{\infty} \phi_n(x_0, y_0, \ldots, x_n, y_n)$, $\forall x_0 \exists y_0 \forall x_1 \exists y_1 \ldots \bigvee_{n=0}^{\infty} \phi_n(x_0, y_0, \ldots, x_n, y_n)$, where the function mapping n to $\phi_n(x_0, y_0, \ldots, x_n, y_n)$ is assumed to be recursive under suitable coding.

We suggest that this manifests a remarkable independence of L from the formalism used, and in that sense provides evidence for Gödel's suggestion that constructibility might be a good candidate for a formalism independent notion of definability – though not in the way he imagined it at the time, evidently. Constructibility being not particularly sensitive to the underlying logic in that sense gives evidence that a type of Church–Turing Thesis holds for the intuitive notion of constructibility. On the other hand there is a clear conflict between logics that L "reads" as first order, in the sense outlined above, and logics that are genuinely first order, i.e satisfying the Lindström characterisation of first order logic, or close to first order logic, again by the Lindström characterisation. The cofinality quantifier (see below) is a case in point: first order logic with the cofinality quantifier adjoined is fully compact, and satisfies the Löwenheim–Skolem Theorem down to $\aleph_1$.[97] However it turns out that L reads cofinality logic as non first order, i.e. one obtains a new inner model with cofinality logic, as we now explain.

4.4.1 The Cofinality Quantifier

The cofinality quantifier introduced by Shelah in [222] says that a given linear order has cofinality ω. The logic is important due to the fact that first order logic with this quantifier adjoined satisfies the compactness theorem irrespective of the cardinality of the vocabulary, and has Löwenhem–Skolem property

---

[95] See [128] for details.

[96] These are quantifiers which say that a given definable equivalence relation has $\aleph_\alpha$ equivalence classes.

[97] Löwenheim–Skolem Theorem down to $\aleph_1$ means that if a sentence of the logic has a model at all, it has a model of cardinality $\leq \aleph_1$.

down to $\aleph_1$. This logic has also a natural complete axiomatisation, provably in ZFC.[98]

The cofinality quantifier Q_κ^{cf} for a regular κ is defined as follows:

$$\mathcal{M} \models Q_\kappa^{\text{cf}} xy\phi(x, y, \vec{a}) \iff \{(c, d) : \mathcal{M} \models \phi(c, d, \vec{a})\}$$

is a linear order of cofinality κ.

A subtle point about this logic is that the inner model arising from the quantifier Q_κ^{cf} *need not witness cofinality* κ; it just "knows" which linear orders have cofinality κ in V, as if the model were equipped with an oracle.

It is surprising that in spite of cofinality logic being very close to first order logic, we have the following result: assuming large cardinals, the inner model obtained from the cofinality quantifier differs from L. Technically: if $0^\sharp$ exists, then $0^\sharp \in L(\mathcal{L}(Q_\kappa^{\text{cf}}))$ for each κ. The model is an interesting one, in that its theory is generically absolute. In fact other models built on generalised quantifiers have been explored in [128], whose theory is also generically absolute, assuming large cardinals.

We can also change our point of view – in essence moving from classes to logics – by considering the equivalence of a logic in the above equivalence relation as a measure of similarity of these logics. Thus logics which are in this sense equivalent to first order logic are then considered "similar to first order logic, as far as constructibility is concerned". Respectively, logics which are in this sense equivalent to second order logic are then considered "similar to second order logic, as far as constructibility is concerned". Do such metrics serve their logics as well as the Lindström characterisation serves first order logic? Should our commitment to first order logic be re-evaluated in the light of these conflicting messages? This is perhaps analogous to the received view of Henkin Second Order logic, which is often seen as simply a many-sorted first order logic – a truism that was undercut by later developments coming from stability theory, showing that there are significant differences between these two logics.[99]

In the above results we considered certain fragments of second order logic. Of these fragments philosophers have been interested mainly in second order

---

[98] The main new axiom is

$$\neg(Qxy\psi(x, y) \wedge Q^*xy\varphi(x, y) \wedge \forall x \exists y \exists v(\theta(v, y) \wedge \varphi(x, y))$$
$$\wedge \forall w \exists x \forall y \forall v((\varphi(x, y) \wedge \theta(v, y)) \rightarrow \psi(w, v))).$$

Here Q is the quantifier "the cofinality is ω" and Q^* is the quantifier "the cofinality is bigger than ω."

[99] Namely, the many-sorted first order theories that come from Henkin second order logic are highly unstable even in the empty vocabulary owing to the Comprehension Axioms, while many-sorted first order theories per se can very well be stable or unstable depending on what kind of theories are considered. Thanks are due to J. Väänänen for this observation.

logic itself, studied extensively in connection with structuralism and other foundational issues. One hopes that the study of its fragments, especially those which are (countably) compact and have completeness theorems,[100] or have other desirable properties, might also be of interest. Here the first order/second order distinction is projected into set theory through the concept of constructibility, with L serving as a first order surrogate and HOD serving as the second order surrogate, i.e. taking HOD in the sense of $L(\mathcal{SO})$, the version of L obtained by replacing the underlying first order logic in the construction of L with second order logic.

4.4.2 Beyond L

It was shown in [128] that in fact the restriction to L is not intrinsic to the analysis. We mentioned that also HOD is not particularly sensitive to the underlying logic and indeed one can apply a concept D of definability to any suitable logic in the operation $\mathcal{L}^* \mapsto D(\mathcal{L}^*)$, and formulate a Church–Turing Thesis of this sort, relative to D.

The version of formalism freeness considered here with respect to definability involves formalism- or more precisely logic-independence, i.e. invariance under substitution of one of a class of logics, considered on a case by case basis. It was crucial to consider not a completely arbitrary class of logics, but rather logics with varying notions of "definable set".

4.4.3 Extending the Axioms of Set Theory

Instead of considering a particular canonical set-theoretic construction, we asked in [128], does the theory ZFC itself, viewed informally, or any of its semantic extensions, admit a Church–Turing Thesis of the kind we have been considering? There are a number of ways of answering the question. Our approach is the following: given a logic $\mathcal{L}^*$, exchange $\mathcal{FO}$ in the Separation and Replacement Axioms of ZFC with another logic $\mathcal{L}^*$, obtaining an $\mathcal{L}^*$-version of ZFC, denoted $ZFC(\mathcal{L}^*)$. More exactly, the modification is that the formula $\phi(x, \vec{y})$ in the Schema of Separation

$$\forall x \forall x_1 \ldots \forall x_n \exists y \forall z (z \in y \leftrightarrow (z \in x \wedge \phi(z, \vec{x})))$$

and the formula $\psi(u, z, \vec{x})$ in the Schema of Replacement

$$\forall x \forall x_1 \ldots \forall x_n (\forall u \forall z \forall z' ((u \in x \wedge \psi(u, z, \vec{x}) \wedge \psi(u, z', \vec{x})) \rightarrow z = z')$$
$$\rightarrow \exists y \forall z (z \in y \leftrightarrow \exists u (u \in x \wedge \psi(u, z, \vec{x})))))$$

---

[100] See [121].

are allowed to be taken from $\mathcal{L}^*$ rather than just $\mathcal{FO}$. We do not add logical rules for $\mathcal{L}^*$-formulas, rather the logic $\mathcal{L}^*$ is *semantically* defined, the concept of a model (M, E), $E \subseteq M \times M$, satisfying the axioms $ZFC(\mathcal{L}^*)$ being obviously well-defined. Our question is, to what extent is $ZFC(\mathcal{L}^*)$ dependent on, or sensitive to, the choice of $\mathcal{L}^*$? Note that $ZFC(\mathcal{L}^*)$ is at least as strong as ZFC in the sense that every model of $ZFC(\mathcal{L}^*)$ is, a fortiori, a model of ZFC. The class of models of ZFC is immensely rich, ZFC being a first order theory. We now ask the question, what can we say about the models of $ZFC(\mathcal{L}^*)$ for various logics $\mathcal{L}^*$?

If we think of the equivalence classes of logics defined by the schema

$$\mathcal{L}^* \equiv \mathcal{L}^{**} \text{ if and only if } ZFC(\mathcal{L}^*) = ZFC(\mathcal{L}^{**})^{101}$$

then the following is true: The class of $\mathcal{L}(Q_0)$ contains all logics that can express "finiteness" and are eliminable[102] in ω-models of set theory, for example weak second order logic. The class of $\mathcal{L}(Q_0^{MM})$ contains all logics capable of expressing well-foundedness, which are eliminable in transitive models of set theory, such as strictly absolute logics, e.g. the recursive game quantifier. Finally, the class of $\mathcal{L}_{\omega_1\omega}$ contains all logics between $\mathcal{L}_{\omega_1\omega}$ and $\mathcal{L}_{\omega_1\omega_1}$. See [128] for details.

It is not hard to see that a model of ZFC is a model of $ZFC(L(Q_0))$ if and only if it is an ω-model. Essentially, this follows from the fact that we can use the quantifier Q_0 to define the standard part of the integers, and hence by induction every integer is standard. An elaboration of this shows that a model of ZFC is a model of $ZFC(L(Q_0^{MM}))$ if and only if it is well-founded [128]. Finally, transitive σ-closed models of ZFC can be similarly characterised.

So similar to the case of $L(\mathcal{L}^*)$, the class of models of $ZFC(\mathcal{L}^*)$ is somewhat dependent on the choice of $\mathcal{L}^*$, while at the same time there is also a great deal of variability as to the choice of $\mathcal{L}^*$.[103]

One can also proceed differently. Matt Kaufmann [118] added the stationary quantifier[104] to ZFC together with some natural axioms concerning the stationary quantifier, and proved in the new system, among other things, the consistency of ZFC. One can also start with arithmetic. Angus Macintyre [156] added the Magidor–Malitz quantifier Q_0^{MM} together with its canonical axioms, to Peano's axioms. James Schmerl and Steve Simpson [214] proved the

---

[101] i.e. $ZFC(\mathcal{L}^*)$ and $ZFC(\mathcal{L}^{**})$ have the same models.

[102] i.e. can be first order defined. The quantifier Q_0 means "there exists infinitely many".

[103] Other extensions of this kind have been considered. See also A. MacIntyre's [156], and Schmerl and Simpson's [214].

[104] [15].

Paris–Harrington sentence in the new system demonstrating its strength over the first order Peano axioms.

4.4.4 Kleene's Ramified Hierarchy

We noted that the calculus we developed in connection with Gödel's constructible hierarchy, leading to the concept of extended constructibility, can be applied to other logical hierarchies – in fact doing so would be very natural. In his [282] P. Welch tests Kleene's Ramified Hierarchy, which is built over second order logic, against various logics, just in the way prescribed by the calculus.

As in Kleene's [133] we define by recursion on α the set $P_\alpha \subseteq \mathcal{P}(\omega)$ as follows:

$$
\begin{aligned}
P_0 &= \varnothing \\
P_\nu &= \bigcup_{\alpha < \nu} P_\alpha, \text{ for } \nu \text{ limit} \\
P_{\alpha+1} &= \{Y \subseteq \omega : Y \text{ definable in } L^2 \text{ over } \langle P_\alpha, \omega, +, \cdot, 0, 1 \rangle\}.
\end{aligned}
$$

Here L^2 refers to second order logic and $\langle P_\alpha, \omega, +, \cdot, 0, 1 \rangle$ refers to the Henkin model in which P_α is the range of second order variables and ω the range of first order quantifiers. Let P be P_α for the least α such that $P_{\alpha+1} = P_\alpha$.

The set P is analogous to Gödel's constructible hierarchy, but for analysis (i.e. second order number theory, or the theory of reals, rather than the theory of sets).

As Welch observes, one can extend L^2 by generalised quantifiers and the set P does not change, for the trivial reason that everything is countable. For example, one can add the quantifier Q_0 (which is non-trivial because we are dealing with Henkin models, not full models), and the cofinality quantifier Q_ω^{cof}. One can also add the uncountability quantifier Q_1. These observations have a trivial proof but they witness the formalism freeness of P, at least of a limited kind, nevertheless. Less trivially, one can add the *game quantifier*

$$
\forall x_0 \exists x_1 \forall x_2 \exists x_3 \ldots R(\langle x_0, x_1, x_3, \ldots \rangle, \vec{y}),
$$

where $R(z, \vec{y})$ is arithmetic (i.e. Σ_n^0, some n) to L^2 and still get the same P.

In this case one obtains something new, i.e. a proper extension of P, if one adds the game quantifier with projective $R(z, \vec{y})$[105] to L^2. Then one obtains the smallest projectively correct model of analysis, projectively correct meaning that definitions of projective sets are absolute between

---

[105] i.e. Σ_n^1 for some n. Alternatively, beginning with closed sets one obtains the projective hierarchy by repeated iterations of taking complements and continuous images.

the model and the standard model of second order arithmetic, assuming Projective Determinacy.[106]

4.5 Logical Autonomy?

Might there be any virtue in foundational practices that routinely display variants instead of routinely choosing among them?[107] The parametric treatment of logics was pursued here for the sake of gaining *logical autonomy* – giving an overview of logics, in this case logics between first and second order, without being entangled in any one of them. But is the Turing analysis of computability (as we have defined it) transferable to the cases of definability and provability, as Gödel suggests in his 1946 lecture? The ingredients of that analysis involve confluence and grounding, where the latter builds on an attunement to what we called the "human profile". We argued that in the case of Gödel's constructible hierarchy L a degree of confluence can be seen by substituting fragments of second order logic for first order definability in the definition of L; and to some degree, the same is true of HOD. The method can be implemented not just for definability in the sense of L or HOD, as was done in [128], but also in other contexts. Kleene's ramified hierarchy of reals (treated by Welch in [282]) is amenable to this treatment as we saw. Other hierarchies can be treated as well, conceivably, with suitable notions of confluence and grounding formulated on a case by case basis.

However we are missing a grounding in the definability case. A Turing analysis in the sense laid out in Chapter 3 denotes an approach that takes a canonical mathematical structure and searches for confluence, and then with confluence being established searches for a grounding of a particular kind. Of course Turing himself simply gave the grounding example, without confluence in the background.

Talk of grounding in connection with the concept of definability in set theory seems premature. That confluence led to grounding so quickly in the case of computability, was, as Gödel said, "a kind of miracle". But the time frame for definability (not to mention the case of provability) in set theory will be very different.[108] Still it would seem that a great deal of confluence is already attached to the two notions of definability Gödel identified, L and HOD.

---

[106] i.e. the game associated to the quantifier is determined if the relation R is projective.

[107] We paraphrase here a passage from Matthew Reynolds et al., Conference Brief, "Prismatic Translation", St. Anne's College Oxford, 1–3 October.

[108] That confluence led to grounding in the computability case is accurate from the point of view of some logicians, but not perhaps all.

Tracking the profile of natural language in foundational practice, which is one way of articulating the project to which this book is dedicated, meant here looking for invariants in frameworks designed for a parametric use of logics. Thus the constructible hierarchy, viewed as a natural (mathematical) object, was tested for structural invariance against certain fragments of second order logic. We found that the Lindström characterisation of first order logic displays anomalies in particular settings; and we found that inner models with large cardinals whose theory is moreover generically absolute can be built by the adjunction of generalised quantifiers. Gödel's search for decidability is a quest for invariance in the form of an intended model of set theory. Such a model should contain large cardinals; and its theory should also be generically absolute as far as possible.

Gödel's methodology involved stepping outside of logical hierarchies, and invoking true large cardinal principles. We followed that path as far as we could, and indeed found some new inner models with desirable properties from the point of view of the inner model programme.

Is there a philosophical moral to be drawn? Our working hypothesis in this area has been that the notions considered here, namely constructibility, hereditary ordinal definability and the informal Zermelo–Frankel axiomatisation with the Axiom of Choice, have semantic content independently of the logic $\mathcal{L}^*$ over which they are built, short of very dramatic variations in the strength of $\mathcal{L}^*$. This is supported by the fact that these notions are to a great degree stable across a variety of distinct formalisms, that is to say, to a degree formalism free. We did not offer a full-blown theory of semantic content here. Instead we focussed our attention on certain hidden regularities and/or anomalies inherent in, as Nadel puts it, our pre-existing definitions, in the modest hope that "any dissatisfaction with the examples presented ... may give some insight into possible changes or improvements in these current notions".[109]

The method given here is reminiscent of Husserl's method of *eidetic variation*. The concept of *eidos*, standing for essence, is a fundamental term for Husserl, especially in *Ideas I*, though the term also appears in the earlier work *Logical Investigations*. In the method of eidetic variation one seeks to grasp an essence through the imaginative free variation within the space of concrete types (of the eidos), beginning with an arbitrarily chosen example. This yields *eidetic seeing*, the seeing of an objectively existing essence "seized upon in its originary mode",[110] as opposed to grasping a manifestation of the essence in the form of a concrete particular.

---

[109] Nadel, [189], p. 103.
[110] [111], p. 341.

Invariance is the key to the discovery of an essence. As Cohen and Moran write in *The Husserl Dictionary*:

> Imaginative free variation plays an essential role in allowing the *eidos* or essence of the phenomenon to manifest itself as the structure of its essential possibilities, what is invariant across all possible variation. In *Phenomenological Psychology*, he [i.e. Husserl JK] gives the example of beginning with a specific shade of red and running though variations until one arrives at the *eidos* red (Hua IX 82). In the *Cartesian Meditations*, Husserl gives the example of seeking the essence of an act of perceiving. Beginning with any current perception, e.g. seeing a table (even one carried out in imagination, i.e. imagining seeing a table), one then seeks to alter the constituent parts of the object, while retaining the perceiving element in the act. The essential features are those which cannot be varied in our imagination.[111]

The space of examples is potentially infinite, so how does one know one has arrived at an essence? One must see that it is "pointless to continue (to keep performing 'and so on')".[112]

Husserlian phenomenology was congenial to Gödel, who may even be said to have converted to the view, at least on the basis of his 1961 lecture manuscript "The modern development of the foundations of mathematics in light of philosophy."[113] As for eidetic language, the language of essences, among the many occasions that Gödel can be seen using it is in his discussions with Wang about the "sharp concept" of mechanical procedure, a concept that was there all along, only its true nature was revealed by Turing.[114]

The framework of extended constructibility could be viewed as a Gödelian framework in the sense of being committed to a sharp informal or intuitive notion, "constructibility in set theory".[115] In the language of eidetic variation we performed a variation within the space of known definability concepts, which were taken to be formal. In some cases we not only saw that it was "pointless to continue", we were able to prove it. For example we found that Gödel's constructible hierarchy is unchanged by substituting any of the quantifiers "there are $\aleph_\alpha$-many", for all α.

The method of eidetic variation presupposes a clean division between essence and instance. Consider for example, the concept of a tone. As Cohen

---

[111] [40], p. 160. Hua IX 82 refers to Husserliana volume IX [277].

[112] ibid.

[113] Published in volume three of Gödel's Collected Works [87], pp. 374–387. See [268] for an extended discussion of Gödel's phenomenological turn.

[114] See Section 3.4.

[115] The presentation of the constructible hierarchy in terms of closure under the Gödel operations, as opposed to the presentation involving first order definability, supports the idea of the constructible hierarchy as given in natural language.

and Moran ask in [40], how can one distinguish between instance and essence in that case? As they remark: "one must know what type the instance falls under in order to vary it to find the essence".[116]

In extended constructibility, the distinction between *essence* and *instance* is clear: this is simply the distinction between constructibility informally construed, as opposed to the various versions of constructibility generated by the different logics considered.

Arana inverts this view of formal languages, though the key distinction here is an epistemological one, between describability and ineffability:

> I suggested thinking about informal language, rather, as the shadow of formal language, or alternately the residue of formal language: the language that we do not express formally but which is still there, not sharply or precisely (thus, in the "dark"), but still visible. It is to think about the informal as what is not formalized, that is, as the complement of the formal. This strategy opposes itself to the usual sense that we all already know what is informal and our task is to build up formal languages from that. Instead, we take the point of view that the formal is what is "easy" to describe and
> the informal isn't.[117]

Our emphasis in this chapter has been on implementation rather than exegesis. Some have thought that the method of eidetic variation was directly inspired by Husserl's thesis (under Weierstrass) on the calculus of variations.[118] In that sense we are perhaps meeting the method at its point of origin.

---

[116] [40], p. 161.
[117] Arana, personal communication.
[118] See [110].

5

Tarski and "the Mathematical"

The suggestion here has been that certain twentieth-century logicians, working under the pressure of various *philosophical* imperatives, were compelled toward natural language, Gödel principally among them. who Gödel had already expressed an interest in informal decision procedures in his 1929 thesis – "informal" meaning here decision procedures given in natural language – asked for absolute, or, on our reading of the term "absolute," formalism independent, notions of definability and provability in his 1946 Princeton Lecture. This led to the idea of hereditary ordinal definability in set theory and more recently to the idea of extended constructibility, giving new inner models of set theory. Formalism independence also became important during key moments in the early history of computability, as we saw. In each of these cases formalism independence was a means to a specific end, a methodology directed toward various aims, e.g. securing decidability for set-theoretical statements (for Gödel) or in the case of computability finding an adequate definition of human effective computability.

Tarski, on the other hand, made a straight run for natural language. Tarski's work in logic is vast and there is accordingly a vast interpretive literature on Tarski's work in logic and set theory, his work on formal semantics, on logicality and on truth. Much of that work was done against the background of the Warsaw school in logic, a dynamic and sophisticated logical milieu to which the work of Tarski bears a complex relationship.[1] Our interest here is in Tarski's conceptualisation of "the mathematical", as he called it, a conceptualisation that grew out of a certain picture of mathematics, and which in turn would contribute mightily to a stream of thought that persisted in model theory from its emergence in the algebraic school in the nineteenth century to

---

[1] We refer the reader to Betti [20], to chapter 8 of Burgess's [31], Consuegra [209], Feferman [57, 61], Frost-Arnold [75], Gómez-Torrente [92–94], Hodges [106], Mancosu [168, 169], Patterson [189], Sher [231], and Woleński and Murawski [186], among others, for an exploration of this complexity.

its reemergence in the work of present day model theorists such as S. Shelah and B. Zilber: work which prioritises the suppression of syntax and logic in one form or another, and the forefronting of semantic concepts.[2] And although "the mathematical" has not figured in the work of contemporary model theorists in the very explicit way it did for Tarski, interestingly enough the concept has reemerged in present day model-theoretic practice – albeit in a more polemic form – especially among model theorists who are committed to first order methodology.

We wish to say at the outset that within the stream of thought in model theory which interests us here, interactions of syntax and semantics are complex. For example the centrepiece of Zilber's work on complex exponentiation is the theorem that it is axiomatisable in $L_{\omega_1,\omega}(Q_1)$.[3]

As for Tarski's own influences, in much of his logical work Tarski himself has owned to working under the very obvious influence of the algebraic tradition in logic today associated with figures like Schröder and Peirce,[4] while others have pointed toward the influence of the successful – in Tarski's view – conceptual analyses of intuitive geometric notions coming out of mainstream mathematics in Poland at the time.[5]

In philosophy, Tarski's published remarks are dwarfed (in quantity) by his logical and set-theoretical work.[6] Nevertheless Tarski did make a number of explicit statements in print over the years. In an early paper, he claimed to identify with the intuitionistic formalism (so-called) of Lésniewski, a view he later disavowed.[7] There is also the emergent finitism and physicalism of the *Wahrheitsbegriff.*[8]

The conversational record begins to pick up in the early 1940s with Carnap's notes of Tarski's conversations with Carnap and Quine at Harvard, revealing Tarski in his strong finitist/nominalist mode. In later years he would refer to himself as a "tortured nominalist"[9] or a "nominalist with some materialistic

---

[2] Semantic in both senses of the term "semantic", arguably, i.e. formal and linguistic, to borrow Burgess's terminology in his [31].

[3] See [289]. Q_1 is the quantifier "there are uncountably many".

[4] The story of formalism freeness could well have begun with the algebraic school. But we are limiting ourselves here to developments after 1930, so picking things up in the middle of the story.

[5] See e.g. [61].

[6] We look forward to the day when such disciplinary binaries as those that are usually drawn between philosophy and related areas in logic and philosophy of mathematics become a thing of the past, but we will not explore that topic here.

[7] [251], p. 62.

[8] See footnote 1, [251], p. 253.

[9] [57], p. 52.

taint".[10] And while a certain "rude anti-Platonism", as he called it, was a mainstay of his evolving philosophical views, at the same time Tarski's philosophical writings and remarks display all the unresolved complexity that comes with being involved in these two disciplines, to wit: philosophy and mathematics. There is the activity of proving theorems, such as: *Let L be a complete lattice and let f : L → L be an order-preserving function. Then the set of fixed points of f in L is also a complete lattice;*[11] and then on the other side of things there is philosophical reflection.

Tarski drew a sharp line between the two. As H. Sinaceur put it:

> [With Tarski JK] We are in front of a new fact in the history of modern mathematical logic: the non-tacit and expressively assumed splitting between logical work as such, on the one hand, and, on the other hand, assumptions or beliefs about the effective or legitimate ways of doing that work and about the nature of the mathematical and logical entities linked with those ways.[12]

As for Tarski's core philosophical views, he obviously had them and even referred to himself as "something of a philosopher".[13] Interpreters seem to be divided between those who see a coherent view in the writings, albeit one . having to be sussed out from a very complex record;[14] and those allowing for a more discontinuous profile, given the possibilities in the record ranging from his interest in nominalism and in developing a finite physicalist language for science, to what appears to be a commitment to semantic realism,[15] if not to the notion of truth *simpliciter*[16] and "the semantic view that truth is not just proof, and meaning not just language".[17]

---

[10] "I represent this very rude kind of anti-Platonism, one thing which I could describe as materialism, or nominalism with some materialistic taint, and it is very difficult for a man to live his whole life with this philosophical attitude, especially if he is a mathematician, especially if for some reasons he has a hobby which is called set theory, and worse – very difficult." Tape recording of remarks given at a joint meeting of the ASL and the APS in Chicago (28–29 April 1965), quoted in [209].

[11] This is known as the Knaster–Tarski theorem.

[12] [238], p. 368. Sinaceur again:
Moreover, he repeatedly claimed he could develop his mathematical and logical investigations without reference to any particular philosophical view concerning the foundations of mathematics. He was eager to disconnect his results from any definite philosophical view, as well as from his personal (varied and variable) leanings. Sinaceur, ibid, p. 367.

[13] In fact it was common among analytic philosophers of the 1980s not to regard Tarski as a member of their profession at all, a view which to some extent persists today.

[14] See e.g. [209], p. 228.

[15] [238], p. 390.

[16] [61], p. 83.

[17] [238], p. 391. See also Murawski and Woleński: "Although all this indicates Tarski's decisive sympathies towards nominalism, reism, and so on, we should note once again the dissonance

Sinaceur sees Tarski's *logical* work as philosophically neutral. Indeed Tarski himself insisted on the neutrality of the semantic conception of truth:

> ... we may accept the semantic conception of truth without giving up any epistemological attitude we may have had; we may remain naïve realists, critical realists, or idealists, empiricists, or metaphysicians – whatever we were before. The semantic conception is completely neutral towards all these issues.[18]

A divided practice, then: proving classical theorems in set theory and model theory while at the same time questioning the axioms that lie at the basis of those theorems, questioning even the idea of infinite set.[19] As Mostowski put it:

> Tarski, in oral discussions indicated his sympathies with nominalism. While he never accepted the "reism" of Tadeusz Kotarbiński, he was certainly attracted to it in the early phase of his work. However, the set-theoretical methods that form the basis of his logical and mathematical studies compel him constantly to use the abstract and general notions that a nominalist seeks to avoid. In the absence of more extensive publications by Tarski on philosophical subjects, this conflict appears to have remained unresolved.[20]

in Tarski's views, namely between his logical and mathematical practice and some of his philosophical views; ... one should study problems using any fruitful methods and making no philosophical presuppositions. There is no need to announce one's philosophical views concerning the investigated problems because this does not belong to scientific duty, this is a 'private' affair. Tarski's attitude was in full accordance with this. To some extent he followed the pattern of doing philosophy in the Lvov-Warsaw School. Twardowski and his students distinguished 'metaphysicism', that is, limiting concrete research by metaphysical assumptions, from genuine scientific work. Although in philosophy this attitude is even more difficult to maintain, if it can be maintained at all, than in mathematics, it had an importance influence on Tarski." [185], p. 38.
And as Patrick Suppes notes, Tarski "was extraordinarily cautious and careful about giving any direct philosophical interpretation of his work" in semantics (1988, 81). ... Frits Staal, a friend and Berkeley colleague of Tarski's, recalls Tarski expressing "[m]ore than once ... that he did not like to talk much about philosophy because he thought it was like giving an 'after dinner' speech – in other words, it was not rigorous". ([57], p. 318). If we take Staal's testimony at face value, it appears that Tarski not only believed his work in semantics to be independent of traditional philosophical concerns, but he also regarded such concerns with some skepticism. [189], p. 225.

[18] [247], p. 362.
[19] Already in the Wahrheitsbegriff, as was noted. From the 1941 conversations between Tarski, Quine and Carnap, quoted in [167], p. 374:

I understand basically only languages which satisfy the following conditions: 1. Finite number of individuals 2. Realistic (Kotarbiński): the individuals are physical things; 3. Non-platonic: there are only variables for individuals (things) not for universals (classes and so on).

[20] [182], p. 81.

What of Tarski's logical work, beyond the semantic conception? A third interpretive approach, and the road we take, is to bracket Tarski's self-account and simply consider the theorems on their own. In that case it would seem that Tarski's logical work is not neutral at all, that in fact he *did* have a position, somewhere in the area of naturalism – or at least the theorems do. Our argument is based on Tarski's conceptualisation of "the mathematical", to which we now turn.

5.1 "The Mathematical", Definable Sets of Reals, and Naive Set Theory

"The mathematical" has something of an odd status in Tarski's writings, being both ubiquitous but at the same time unanalysed – not unlike the status of, for Tarski, "the semantical", notwithstanding the various physicalist interpretations of the latter.[21] This is in comparison to, for example, "the logical", a concept that preoccupied Tarski for obvious reasons having to do with the status of logical constants relative to the notion of semantic consequence.

Of course the meaning of "the mathematical" comes through in the theorems – it is part of *their* philosophy. For example in his 1931 paper "On definable sets of real numbers" [245] (henceforth DSRN) "the mathematical " is identified with naive set theory, i.e. with the Boolean operations, and projection.[22] That said, Tarski did make a few remarks on the meaning of the mathematical, as did Vaught in the quote directly below, and also Carnap, who recorded a few stray remarks Tarski made on the topic during conversations in the 1940s at Harvard with Carnap and Quine.

Before turning to Tarski's writings on the mathematical here is Vaught in 1986 [274], commenting on Tarski's 1950 ICM Lecture:

> An additional feature is that in the whole presentation Tarski (returning to and expanding his old methods from [245]) manages to define notions like "**EC** class" without any mention of a formalized language. Tarski liked the idea of replacing a "metamathematical" definition by a "mathematical" one; and was even more pleased by a "very mathematical" one such as Birkhoff's definition [21] of equational class. Later on he very much liked the "purely mathematical" definition of $\mathfrak{A} \equiv \mathfrak{B}$ by R.Fraïssé [elementary equivalence in terms of EF games: JK, [73]],

---

[21] For more on the relation between the physicalism and the semantic conception of truth see Section 5.2.

[22] Henceforth we will usually drop quotation marks from "the mathematical," for ease of reading.

and still later the definition using ultraproducts (see below) [of Keisler/Shelah: [120], [220] JK]. These very suggestive intuitive ideas may be without a precise content, as a precise distinction between "mathematical" and "metamathematical" might well be considered to be impossible because of Tarski's definition of truth! Of course it is only in proofs that mathematicians must be precise. In the important matter of selecting what to think about, anything goes![23]

Vaught's remark that Tarski's truth definition is a device for dissolving the distinction between "the mathematical" and "the metamathematical" is tantalising. As for Vaught's examples, relevant here is the fact that Birkhoff's theorem characterises an equational class by means of set-theoretical closure operations, together with the notion of homomorphism. Precisely: a class of structures is an equational class if and only if it is closed under subalgebra, direct product and homomorphic images. As for the *EF* or Ehrenfeucht–Fraïssé games, we explored the idea of a logic being given by a game in Section 2.2.3, which included the above characterisation of elementary equivalence.

Turning now to Tarski's own writings, his 1931 paper "On Definable Sets of Real Numbers" [245] seems to contain the first mention of the mathematical in Tarski's work. DSRN is of course known for introducing the notion of satisfaction, albeit in a somewhat preliminary form, though not the notion of satisfaction in a domain, a crucial point.[24] Tarski opens the paper thus:

> Mathematicians, in general, do not like to deal with the notion of definability; their attitude toward this notion is one of distrust and reserve. The reasons for this aversion are quite clear and understandable. To begin with, the meaning of the term 'definable' is not unambiguous: whether a given notion is definable depends on the deductive system in which it is studied … It is thus possible to use the notion of definability only in a relative sense. This fact has often been neglected in mathematical considerations and has been the source of numerous contradictions, of which the classical example is furnished by the well-known antinomy of Richard. The distrust of mathematicians towards the notion in question is reinforced by the current opinion that this notion is outside the proper limits of mathematics altogether … In this article I shall try to convince the reader that the opinion just mentioned is not altogether correct. Without doubt the notion of

---

[23] [274], p. 875. Errata in [273].

[24] As Feferman writes:

> In fact, the general notion of structure for a first-order language L is already described in Tarski's 1950 ICM address, essentially as follows: a structure A is a sequence consisting of a non-empty domain A of objects together with an assignment to each basic relation, operation, and constant symbol of L of a corresponding relation between elements of A, operation on elements of A, or member of A, resp.

> [61], p. 75. The first time the mathematical definition of truth in a structure appears in print is dated by W. Hodges to Tarski and Vaught's 1957 [254].

definability as usually conceived is of a metamathematical origin. I believe that I have found a general method which allows us to construct a rigorous metamathematical definition of this notion. Moreover, by analysing the definition thus obtained it proves to be possible (with some reservations to be discussed at the end of § 1) to replace it by a definition formulated exclusively in mathematical terms. *Under this new definition the notion of definability does not differ from other mathematical notions and need not arouse either fears or doubts; it can be discussed entirely within the domain of normal mathematical reasoning.*[25]

The rigorous metamathematical definition Tarski refers to is the standard logical notion, given in the *Wahrheitsbegriff*,[26] and more exactly in [254]. Here he limits himself to "definable set of reals", definable relative to the first order theory of real closed fields, built over simple type theory. The point here is that the *mathematical* definition of definability is effected by replacing the notion of formula, or in his terminology, sentential function, with, again in his terminology, *its mathematical analogue* – namely the set of sequences that satisfy the function. Tarski notes that there can be no global definition of definability, i.e. handling sentential functions of all orders simultaneously, due to Richard's paradox. And he hints at a decision procedure in the case of the theory of real closed fields, though the proof of this via quantifier elimination was given only in his 1951 paper "A Decision Method for Elementary Algebra and Geometry" [248].

There are actually two notions of definability given in the paper, a general notion designed to handle arbitrary interpreted languages, and a definition proceeding by elimination of quantifiers. The second definition yields the following

Theorem 5.1.1 *A set of reals is definable*[27] *if and only if it is the union of a finite number of intervals with rational endpoints.*

The definable sets in the first general case are generated by the satisfying sequences corresponding to the atomic formulas, closing under the standard Boolean operations together with projection.[28] In Tarski's view this assimilates

---

[25] [251], p. 110, emphasis ours.

[26] Footnote 1, page 194 of [251]. The German translation from the original Polish version of Tarski's paper is entitled "Der Wahrheitsbegriff in den formalisierten Sprachen" [246] and appeared in 1935. The English translation is entitled "The concept of truth in formalized languages" [251] and appeared in 1956. Here and subsequently by "Wahrheitsbegriff" we mean to refer to the English version.

[27] in the first order theory of real closed fields.

[28] Kunen makes a similar move in the opening lines of the chapter of his classic text [146] entitled "Defining definability," at which point he remarks that "This approach brings out the fact that the notion of definability does not depend on the specific syntactical details of a development of first order logic."

the metamathematical notion of definability in general to set theory – or, if you like, to "the mathematical" – by means of replacing the syntactic notion of sentential function by what is, again in Tarski's view, a natural set-theoretic object, namely the set of sequences satisfying it – a move that would be replicated again and again in Tarski's future work.

Feferman sees Tarski here as appropriating the set-theoretical methods of the Polish topologists, whose use of set theory in conceptual analysis was, for Tarski, paradigmatic.[29] Feferman himself was critical of the move, remarking that "Tarski assimilated logic to higher set theory to (what I regard as) an unjustified extent."[30] And while we would argue with the phrase "*higher* set theory," Feferman is pointing out a crucial aspect of Tarski's overall methodology, namely the identification of what one would nowadays refer to as naive set theory, the basic axiomatic theory providing the mathematical tools for forming e.g. sums and products of structures, with "the mathematical". Again from Tarski:

> The idea of the reconstruction is quite simple in principle. We notice that every sentential function determines the set of all finite sequences which satisfy it. Consequently, in the place of the metamathematical notion of a sentential function, we can make use of its mathematical analogue, the concept of a set of sequences.[31]

Or as Feferman puts it again:

> However, he may have thought that he could not make clear to mathematicians that these results were part of mathematics until he showed how all the logical notions involved could be defined in precise mathematical (*i.e. general set-theoretical*) terms.[32]

As for the method by elimination of quantifiers, Tarski notes at the end of the paper that this may give rise to the assimilation of syntax to *geometry*:

> In particular, the operation $\sum_k^0$ is easily recognized as that of projection parallel to the X_k, axis ... Ths.1 and 2 take the form of certain theorems of analytic geometry ... The joint note by Kuratowski and myself, which immediately follows this article, as well as further investigations by Kuratowski, seem to testify to the great heuristic importance of the geometrical interpretation of the constructions sketched in this article.[33]

[29] [61], p. 75.
[30] [61], ibid.
[31] [251], p. 120.
[32] [61], p. 80. Emphasis ours.
[33] ibid, p. 141.

This normalisation of metamathematical concepts via geometry and in particular via the notion of dimension would culminate many years later in the Main Gap Theorem of S. Shelah, in which one attaches a notion of dimension to so-called "classifiable" first order theories – normalising, as it were, the notion of first order theory.[34] The conversion of the metamathematical to the geometric also comes to fruition in the work of the model theorist B. Zilber, in which the syntax/semantics distinction altogether is recast in terms of varieties (syntax) and curves (semantics), as was noted in Chapter 1.[35] And then again in Tarski's own work one sees a very explicit geometric inspiration behind his analysis of logicality.[36]

The move to assimilate metamathematics to mathematics – *the mathematical at this point being conceived in terms of naive set theory* – would be repeated in Tarski's logical work going forward. What is interesting about DSRN is that it was written before the set-theoretic paradigm became entrenched in Tarski's work.[37] This fact, namely that the metamathematical definition of definable set of reals is given in DSRN in terms of simple type theory, may give power to Tarski's conflation of naive set theory with the mathematical, for the fairly obvious and simple reason that the metamathematical definition was *not* given in set-theoretical terms.

By the mid-1930s set theory begins to vibrate in different registers in Tarski's work: as a formal theory on the metamathematical side, and as an informal theory on the mathematical side. In conversations between Tarski, Quine and Carnap at Harvard in 1941, we know from Carnap's notes on these conversations that set theory was considered in its formal mode, at least it appears to have been in connection with the conversations between the three on the problem of a finitistic language for science. Part of those conversations were devoted to the question whether set theory belonged to logic or not. Tarski's remark that "mathematics = logic + $\in$" would seem to indicate that

---

[34] Shelah's Main Gap Theorem divides all countable first order theories into two categories: in the classifiable case, there is a bound on the number of models (up to isomorphism), and they can be characterised by a tree of geometric invariants, like the dimension of a vector space. While at the same time in the non-classifiable case, there is a precise sense in which no notion of dimension can be extracted, and the case is chaotic in the sense that the structures are hard to tell apart. [223], and see below.

[35] See [112].

[36] This uses concepts such as the automorphism group attached to the underlying domain. See Section 5.4.

[37] According to Hodges the transition in Tarski's early work from simple type theory to informal set theory is completed by early to mid-1930s. As Hodges puts it: "The deductive theories in question (such as RCF) are formulated in simple type theory; by 1935 the axioms for RCF are regarded as a definition within set theory." [106], p. 118.

set theory was considered in its informal mode. Given the context of those conversations, this seems odd.[38]

The ambiguation would persist. Thus formalised set theory is considered in its foundational role in the *Wahrheitsbegriff*:

> In order to give the following exposition a completely precise, concrete, and also sufficiently general form, it would suffice if we choose, as the object of investigation, the language of some one complete system of mathematical logic. Such a language can be regarded as a universal language, in the sense that all other formalized languages – apart from 'calligraphical' differences – are either fragments of it, or can be obtained from it or from its fragments by adding certain constants ... As such a language we could choose the language of the general theory of sets which will be discussed in §5 ...[39]

While on the other hand, in his 1946 Princeton Bicentennial remarks Tarski distinguished "set theory itself" from logic:

> In discussing the relations between set theory itself and logic I shall begin by saying that I believe that the set-theorist may expect much from the formal logician. I believe that certain problems of set theory may actually be independent of the axioms of set theory and may be shown to be so independent by formal logical means.[40]

Finally, in the 1965 International Colloquium in the Philosophy of Science, held at Bedford College, London, Tarski is recorded as saying:

> The fact that we use set theoretical notions in metamathematical or metalogical discussions seems to me very important, because in such discussions we use the word 'set' in a non-formalized way. Thus there is a difference between set theory and other kinds of mathematical theories.[41]

---

[38] See Mancosu's "Harvard 1940–1941: Tarski, Carnap and Quine on a finitistic language of mathematics for science," [167].

[39] [246], reprinted in [251]. p. 210, footnote 2.

[40] [238].

[41] Consuegra, [209] p. 256. Consegrua notes that in 1940 Tarski seemed to have aspired to a system without set theory, if what he describes below can indeed be done without set theory: Perhaps something completely different will develop. It would be a wish and perhaps a guess, that the whole general set theory [die ganze allgemeine Mengenlehre], as beautiful as it is, will disappear in the future. Platonism begins with the higher levels. The tendencies of Chwistek and others ('Nominalism') to talk only about designatable things [Bezeichenbarem] are healthy. The only problem is finding a good execution. Perhaps roughly of this kind: in the first language, natural numbers as individuals, as in Language I, but perhaps with unrestricted operators; in the second language, individuals that are identical with or correspond to propositional functions in the first language, thus to the properties of natural numbers expressible in the first language; in the third language, the properties

A brief digression about the phrase "naive set theory". We have argued that Tarski conflated naive or informal set theory with the mathematical in DSRN and elsewhere. Mostowski, Tarski's first student, seems to have been the first to coin the phrase in his 1939 paper "On the independence of the well-ordering theorem from the ordering principle" [183]:

> In the system 6 a considerable part of the "naive" set theory can be reconstructed, namely that part in which only sets of power below the first unattainable aleph are considered.[42]

Mostowski goes on to be more specific, explicitly identifying naive set theory with natural language, although the phrase "natural language" is not used here:

> Since it would be very cumbersome to express the proof in the formalism of von Neumann's system, we hold to the language of naive set theory, familiar to all set theoreticians. The complete translation of the proof into the framework of von Neumann's system presents no essential difficulties. This explains the fact why in our exposition we can afford to use several notions which are of imprecise or even contradictory character in naive set theory but are admissible in von Neumann's system. To such notions belongs, for example, the notion of the domain of all sets or of all ordinal numbers. The notion of ordinal number, which is often not sufficiently precise in naive set theory, should be understood in the sense given by von Neumann.[43]

By 1966, Mostowski drops the quotation marks on "naive", so that now "naive set theory" simply refers to Cantor's unformalised theory:

> The Zermelo-Fraenkel (and Bernays-Gödel) set theory arose from attempts to formulate in a consistent way the intuitive assumptions of the naive (Cantorian) set theory.[44]

H. Wang, citing von Neumann, is perhaps more explicit in defining naive or intuitive set theory as including absolute (unrestricted) comprehension:

---

expressible in the second language are taken as individuals, and so forth. So in each language one has only variables for individuals, but nevertheless they cover entities of different levels. Quoted in [209], p. 238.

[42] [183], p. 291.

[43] ibid, pp. 292–292.

[44] Mostowski, *Thirty Years of Foundational Studies*, [181], p. 138.

> The basic intuitive concept is often called naive set theory and identified with a
> belief in an absolute comprehension principle according to which any property
> defines a set.[45]

This is in contrast to Mostowski's formulation, which can be construed both
formally and informally.[46]

The use of the term "naive set theory" has continued in the post-Mostowski
and/or the contemporary period of model-theoretic practice. This is as opposed
to a practice viewing model theory as embedded in a formal set-theoretical
framework. For example, D. Kueker's remark about Abstract Elementary
Classes: "Note that this definition is purely set-theoretic. In particular there
is no syntax ... " is firmly in the spirit of Mostowski's usage.[47] And L. van den
Dries, echoing Tarski's 1941 definition of mathematics as "logic + ∈",[48] opens
his 2016 lecture notes on model theory thus:

> Also, their [i.e. Bolzano, Boole, Cantor, Dedekind, Frege, Peano, C.S. Peirce, and
> E. Schröder JK] activity led to the view that logic + set theory can serve as a basis
> for all of mathematics.[49]

Van den Dries goes on to recommend Halmos's *Naive Set Theory* [96] to his
students, a text somewhat notorious for its view of set theory, though not for
its informal but exact presentation of it.[50]

Model theorists, then – or at least some of them – seem to draw a clear
distinction between naive set theory – their metatheory of choice – and formal
(ZFC) set theory. What does this say about model-theoretic practice? Viewed
as a fully *universal* language, universal in the sense of the *Wahrheitsbegriff*,
natural language admits antinomies. This is noted by Tarski in that text as

---

[45] [278], p. 191.

[46] In his review of Tarski's [250], in which a structural characterisation of universally
axiomatised theories is presented, Abraham Robinson describes the closure properties of the
characterisation as "set-theoretical". As Robinson says, "This series of papers is concerned
with the relations between the syntactical properties of given sentences (for the most part, in
the lower predicate calculus) and the set-theoretical properties of the classes of models
defined by them."

[47] [145].

[48] See [167], p. 366.

[49] https://faculty.math.illinois.edu/ vddries/main.pdf, pp. 1–2.

[50] From van den Dries's text:

> We shall use this section as an opportunity to fix notations and terminologies that are used
> throughout these notes, and throughout mathematics. In a few places we shall need more set
> theory than we introduce here, for example, ordinals and cardinals. The following little book
> is a good place to read about these matters. (It also contains an axiomatic treatment of set
> theory starting from scratch.)

establishing the need for the theory/metatheory distinction and the idea that
truth for a language must be defined in the metalanguage (etc.):

> A characteristic feature of colloquial language (in contrast to various scientific
> languages) is its universality. It would not be in harmony with the spirit of this
> language if in some other language a word occurred which could not be translated
> into it; it could be claimed that 'if we can speak meaningfully about anything at all,
> we can also speak about it in colloquial language'. If we are to maintain this
> universality of everyday language in connexion with semantical investigations, we
> must, to be consistent, admit into the language, in addition to its sentences and
> other expressions, also the names of these sentences and expressions, and sentences
> containing these names, as well as such semantic expressions as 'true sentence',
> 'name', 'denote', etc. But it is presumably just this universality of everyday
> language which is the primary source of all semantical antinomies, like the
> antinomies of the liar or of heterological words. These antinomies seem to provide
> a proof that every language which is universal in the above sense, and for which the
> normal laws of logic hold, must be inconsistent.[51]

Model theorists working in a naive set-theoretic metalanguage must then
choose between ignoring worries about working in an inconsistent language;
or on the other hand they may challenge the idea that natural language is
fully universal.[52] In fact there is every reason to think that Tarski (in the
Wahrheitsbegriff) sees the natural language concept of truth as inconsistent
tout court:

> If these observations are correct, then the very possibility of a consistent use of the
> expression 'true sentence' which is in harmony with the laws of logic and the spirit
> of everyday language seems to be very questionable, and consequently the same
> doubt attaches to the possibility of constructing a correct definition of this
> expression.[53]

As an aside, in an interesting footnote appearing late in the paper Tarski
ruminates, not about the behaviour of the truth predicate in natural language,
but about its finitistic nature:

> In the course of our investigation we have repeatedly encountered similar
> phenomena: the impossibility of grasping the simultaneous dependence between
> objects which belong to infinitely many semantical categories; the lack of terms of
> 'infinite order'; the impossibility of including in *one* process of definition, infinitely

---

[51] [251], p. 164.
[52] I have encountered both positions among working model theorists.
[53] [251], p. 165.

many concepts and so on .... I do not believe that these phenomena can be viewed
as a symptom of the formal incompleteness of the actually existing languages –
their cause is to be sought rather in the nature of language itself: language, which is
a product of human activity, necessarily possesses a 'finitistic' character, and
cannot serve as an adequate tool for the investigation of facts, or for the
construction of concepts, of an eminently 'infinitistic' character.[54]

Tarski seems to be saying here that formal languages are in some sense *parasitic* off natural language(s): they do not introduce new conceptual resources, or have them on their own, over and above those already found in natural language – which is itself limited in important ways.

Returning to the consistency problem, the logical terrain here is subtle. In DSRN Tarski interprets the theory explicitly, i.e. the structure in question is simply taken to be the real numbers. A metatheoretic dilemma can emerge here, if we let it: if we regard the formal definition as being embedded in formal first order set theory we ruin the picture of an interpreted practice, due to a proliferation of interpretations. On the other hand if we regard ourselves as working in natural language – the medium from which the interpretation is drawn – then we are wading into inconsistency.

What about the term metamathematics? Tarski's definition of *metamathematics* seems to have been entirely the standard one.[55] Metamathematics studies "formalized deductive disciplines ... in roughly in the same sense in which spatial entities form the field of research in geometry". As examples of metamathematical concepts Tarski cited, as one would expect, consistency and completeness.[56] In his 1950 paper "Some notions and methods on the borderline of algebra and metamathematics" [249] Tarski will simply equate "metamathematical" with "syntactical".[57]

As for the precise meaning of the mathematical, this is shown but not said, i.e. the theorems (of DSRN) are meant to provide the meaning of the term.[58] One notes that Tarski's conceptualisation of the mathematical draws no distinction between syntax and semantics. Thus "the mathematical" is an formalism

---

[54] [251], p. 253.

[55] Although in 1930 he would define metamathematics quite broadly, as "Concepts having to do with the methodology of the deductive sciences, following Hilbert." As Vaught would comment in his [273], "Today this terminology seems odd."

[56] Quotation from Tarski, "On some fundamental concepts of metamathematics," 1930 (FCM) [251], p. 30.

[57] "This extension is immediate once the completeness theorem for real closed fields has been translated into the language of arithmetic classes; however, it could hardly be derived in a purely metamathematical (syntactical) way from the completeness theorem itself – unless we allow ourselves to apply some rather intricate semantical notions and methods." [249], p. 719.

[58] As Baldwin notes, "Tarski takes his readers to understand the notion of 'mathematical property'." [10], p. 294.

free concept, also in a more restricted sense of formalism freeness, govern-
ing mathematical discourse in which no distinction is drawn between syntax
and semantics, as is the case here. Tarski does consider the question in DSRN
whether a mathematical definition is *adequate*, that is in the case of definability "the question ... whether the definitions just constructed [in DSRN JK]
... are also adequate materially; in other words do they in fact grasp the current
meaning of the notion as it is known intuitively?" Tarski waves the question
away – it is not a mathematical question. One can prove the validity of the
reconstruction by induction, in metamathematics. However within the domain
of mathematics:

> If we wish to convince ourselves of the material adequacy of Def. 10 and of its
> conformity with intuition without going beyond the domain of strictly
> mathematical considerations, we must have recourse to the empirical method.[59]

In his essay "Tarski's Theory of Definition" Hodges examines Tarski's various uses of the term "empirical", concluding that "The examples don't all fit
any obvious pattern, but in several of them Tarski seems to be describing an
informal deduction."[60] There is a method Hodges calls "Sample", which he
traces to Walter Burley's logic text of the late 1320s called "On the Purity
of the Art of Logic". "Sample" is a way of reaching a universally quantified
conclusion:

> Sample: To prove that all A's are B's, look at a suitable sample of A's and check
> that all A's in the sample are B's. (This method can lead to false results, for
> example if through lack of imagination you miss an important sort of A.)[61]

For Hodges, "sample" is "a kind of argument that exactly fits Tarski's
description when one looks back to the days before logicians had proof by
induction".[62]

Suppose we want to eliminate the existential quantifier in $\exists x \phi(x, \vec{y})$, where
$\phi(x, \vec{y})$ is a quantifier-free formula in the theory of real closed fields. The
procedure is as follows:

- Put $\phi(x, \vec{y})$ into disjunctive normal form

$$\bigvee_i \bigwedge_j \psi_{i,j}(x, \vec{y}),$$

[59] [251, p. 129].
[60] [106], p. 98.
[61] [105].
[62] Hodges, ibid.

where each $\psi_{i,j}$ is atomic or negated atomic.

- Note that atomic means now an equation as there are no relations, as inequalities $t < t'$ can be replaced by $\exists x (x \neq 0 \wedge t + x^2 = t')$, and $x \neq 0$ can be replaced by $\exists y (y \cdot x = 1)$.
- Thus

$$\exists x \phi(x, \vec{y}) \leftrightarrow \exists x \bigvee_i \bigwedge_j \psi_{i,j}(x, \vec{y}) \leftrightarrow \bigvee_i \exists x \bigwedge_j \psi_{i,j}(x, \vec{y}).$$

Note that it now suffices to eliminate the existential quantifier from $\exists x \bigwedge_j \psi_{i,j}(x, \vec{y})$, where each $\psi_{i,j}(x, \vec{y})$ is of the form $t = t'$. Since $t = t' \wedge u = u'$ can be replaced by $(t - t')^2 + (u - u')^2 = 0$, we are left with eliminating the existential quantifier from $\exists x (t(x, \vec{y}) = 0)$, where t is a polynomial.

As a concrete example – a very simple one in order to illustrate the basic idea – consider: $\exists x (y x^2 + 2x - 5 = 0)$. We want to eliminate $\exists x$.

- $\exists x (y x^2 + 2x - 5 = 0)$ is certainly true if $y = 0$. So assume $y \neq 0$.
- Then

$$y x^2 + 2x - 5 = 0 \iff x^2 + \frac{2x}{y} - \frac{5}{y} = 0 \iff$$

$$(x + \frac{1}{y})^2 - \frac{1}{y^2} - \frac{5}{y} = 0 \iff (x + \frac{1}{y})^2 = \frac{5}{y} + \frac{1}{y^2}.$$

A solution exists iff $5/y + (1/y)^2 \geq 0$ iff $y \geq -1/5$.
- Thus: $\exists x (y x^2 + 2x - 5 = 0)$ iff $y \geq -1/5$.

That is "sample". The general case is similar: one transforms the logical form of the statement, and then one applies analytic methods given by Sturm's Theorem.[63]

This sets the metamathematical proof on the same explanatory level as the empirical method, which is carried out in natural language. Of course "sample" depends on standard mathematical methods for its validity – a view which is very much on display throughout DSRN. We will return to this point below.

5.1.1 Tarski after DSRN

In his 1950 "Some notions and methods on the borderline of algebra and metamathematics" [249] Tarski revisits the idea of normalising definability:

---

[63] For the statement and proof of Sturm's Theorem see [271], p. 220.

The notion of an arithmetical class is of a metamathematical origin; whether or not a set of algebraic systems is an arithmetical class depends upon the form in which its definition can be expressed ...

However, it has proved to be possible to characterize this notion in purely mathematical terms and to discuss it by means of normal mathematical methods. The theory of arithmetical classes has thus become a mathematical theory in the usual sense of this term, and in fact it can be regarded as a chapter of universal algebra.[64]

So first order definability in general is considered, not just for sets of real numbers. Except that here Tarski algebraises (or mathematises) first order definability by characterising elementary classes through the notion of a so-called cylindric algebra. In terms of the earlier construction, projection is replaced by cylindrification, i.e. existential quantification is expressed by unary operators (cylindrifications).

The [mathematical] theory concerns arbitrary algebraic systems formed by a nonempty set A, some operations $O_1, \ldots O_j$, under which A is closed, some relations $R_0, \ldots R_i$ between elements of A, and possibly some distinguished elements $c_0, c_1, \ldots$ of A.[65]

One considers the family of all elementary classes, defined syntactically. One then observes that the family has certain mathematical properties (e.g. the elementary classes form a base for a topology), i.e. one "forgets" the first order logic. Next one defines mathematically some simple elementary classes from identity and atomic relations. By closing this family under Boolean operations and cylindrification one obtains exactly the original family of all elementary (i.e. first order definable) classes. The Borderline paper traces out a *squeezing argument*, arguably, for the concept of first order definability (see Section 5.3).

The Borderline paper also includes a rephrase of the Löwenheim–Skolem Theorem:

---

[64] [249], p. 705. An arithmetical class is a class of structures definable in first order logic with equality. In Tarski's later (and in the current) terminology this is referred to as an elementary or EC class.

[65] Incidentally we see here what appears to be the first mention of the concept of "structure", especially considering footnote 2:

As is known, we could restrict ourselves without loss of generality to systems formed by a set and certain relations between its elements.

Tarski and Vaught's [254] of 1956 gives the semantics for formal languages in definitive set-theoretical terms, i.e. a predicate for the domain is added to the metatheory.

Theorem 5.1.2 *For every algebra $\mathfrak{A} \in \mathcal{A}$ of an infinite power α and for every infinite cardinal β there is an algebra $\mathfrak{B} \in \mathcal{A}$, of the power β such that $\mathfrak{A} \equiv \mathfrak{B}$. If in addition $\beta \leq \alpha$, then such an algebra $\mathfrak{B}$ can be found among the subalgebras of $\mathcal{A}$.*[66]

Tarski also gives an algebraic version of the compactness theorem, stated without proof.

Other examples of Tarski's programme to algebraise metamathematics: The characterisation of universally axiomatised theories in [250] in terms of closure properties of a class of structures: A class of structures in a finite relational language is universally (or $\forall_1$) axiomatisable if and only if it is closed under isomorphism and substructures, and if for every finite sub-structure B of a structure A, $B \in K$, then $A \in K$.

These results would serve as templates for future model-theoretic work, in particular the notion of an abstract elementary class due to Shelah is given by structural properties of a similar kind.[67] The essential idea throughout is treating definability in a logic, or more precisely, model classes which are definable in a logic (either in first order logic or in a stronger logic), autonomously, i.e. mathematically or algebraically, so "forgetting" or suppressing the logic.

As to the question whether, at the end of the day, it is really possible to "forget" or suppress the logic, this will ultimately depend on one's point of view. Some take the view that syntax is always present in one way or another. For example for J. Barwise, the primary object of study in model theory "is the relationship between syntactic objects and the structures they define". And J. Baldwin has invoked the concept of an implicit syntax.[68]

We will return to the idea of "forgetting the logic" below. As for the idea of "implicit syntax" it was mentioned in section 6.1 and we take it up in Section 6.2 in the form of the broader question, whether and under which circumstances a (or "the") syntax (or logic) can be read off a particular semantic framework.

---

[66] "On the other hand, Theorem 23 can easily be recognised as a mathematical translation of a familiar metamathematical result, in fact, of an extension of the Löwenheim-Skolem theorem." [251], p. 712.

[67] See Section 6.1.

[68] The phrase seems to have appeared for the first time in this context in Baldwin's [10], footnote 19: "Shelah's presentation theorem, discussed below, shows there is an 'implicit' syntax." See Section 6.1 for a further discussion of implicit syntax.

5.2 Tarski's Naturalism

On the surface Tarski's programme to normalise metamathematics represents, oddly, a reversal of logicism. At the same time we have this in *favour* of logicism, from Tarski's 1941 *Introduction to Logic and to the Methodology of the Deductive Sciences*:

> the ... fact that it has been possible to develop the whole of arithmetic, including the disciplines erected upon it – algebra, analysis, and so on – as a part of pure logic, constitutes one of the most remarkable achievements of logical investigations.[69]

One must allow for the idea that a surface discontinuity may mask an underlying connected field. Rather than attempt to resolve the disunities, as it were, of Tarski's logico-philosophical practice, we take the approach of bracketing Tarski's self-account and surveying that practice from the standpoint of the theorems. It then seems obvious that Tarski did take a definite view, one which endowed mathematics with a special status, namely, the status of being *beyond sceptical critique*.

Hilbert expressed the same thought, writing in 1922 of mathematics' "old reputation of incontestable truth".[70] Of course there is a range of interpretations of what Hilbert has said, but this is the Hilbert of C. Franks's [74], to whom Franks attributes "a pre-theoretic belief in the consistency of mathematics that on the one hand does not ride on the promise of a realization of his foundational program, and on the other hand carries on unscathed by announcements of skepticism from the philosophical schools".[71]

Franks argues that the Hilbert programme was a direct implementation of this view of mathematics:

> This is because the question inspiring him [i.e. Hilbert JK] to foundational research is not whether mathematics is consistent, but rather whether or not mathematics can stand on its own – no more in need of philosophically loaded defense than endangered by philosophically loaded skepticism. All the traditional "Hilbertian theses" – formalism, finitism, the essential role of a special proof of consistency – are methodological principles necessitated by this one question. When they are understood in that light, they appear no longer to be the glib scientistic principles of an expert mathematician's amateur dabbling in philosophy. They appear rather

---

[69] [253], p. 76.
[70] [1922] Hilbert, D. Neubergründung der Mathematik. Erste Mitteilung, Abhandlugen aus dem Mathimatischen Seminar der Hamburgischen Universität 1: 157–77. Translated by W. Ewald as "The new grounding of mathematics: first report" in Mancosu [166], pp. 198–214.
[71] [74], p. 2.

to be the constraints on method needed for probing a deep epistemological issue left untouched by rival programs.[72]

The motivation behind Tarski's lifelong programme to "normalise" meta-mathematics for the mathematician is of course different in character. But like Hilbert (or Franks's Hilbert at least), whatever the actual contours of Tarski's actual view was, Tarski's faith in the groundedness, stability and coherence of the natural language practice of mathematics, albeit more or less unstated, is very clear. For what would be the point otherwise, of expressing metamathematical concepts in mathematical terms, if *the mathematical* was set in any way on shaky ground?

And what more can we say about Tarski's faith in the natural language practice of mathematics? For Frege natural language carries something like ontological force, in that a singular term in natural language, so *not* a formal language but a natural language, denotes an object, for Frege, tout court. Moltmann has challenged this notion recently:

> Natural language presents a very different view of the ontological status of natural numbers. On this view, numbers are not primarily abstract objects, but rather 'aspects' of pluralities of ordinary objects, namely number tropes, a view that in fact appears to have been the Aristotelian view of numbers. Natural language moreover provides support for another view of the ontological status of numbers, on which natural numbers do not act as entities, but rather have the status of plural properties, the meaning of numerals when acting like adjectives. This view matches contemporary approaches in the philosophy of mathematics of what Dummett called the *Adjectival Strategy*, the view on which number terms in arithmetical sentences are not terms referring to numbers, but rather make contributions to generalizations about ordinary (and possible) objects. It is only with complex expressions somewhat at the periphery of language that reference to pure numbers is permitted.[73]

Of course one will not see anything like these views spelled out in Tarski's writings, especially those owning to various forms of nominalism.

As for the philosophical basis of Tarski's work in *semantics* it has been suggested that the truth definition was motivated by a belief in physicalism, the idea that semantic facts are always reducible to physical facts. The suggestion was first put forward by Field in his 1972 [65], and elaborated by many others, so that (according to some) it is now the received view – in fact all that remains open for debate in the literature on the physicalistic basis of the truth definition

---

[72] [74], p. 3.
[73] [177], p. 2.

is the question whether and to what degree Tarski's truth definition achieves his purportedly physicalist aims – if Tarski achieves them at all.

However as Frost-Arnold points out in [75], the textual evidence for this is minimal: one sentence in the introduction to the *Wahrheitsbegriff*.[74] As for the reason Tarski mentions physicalism at all, Frost-Arnold explains it this way:

> Seen in this light, perhaps Tarski mentions physicalism not because it is a fundamental conviction that drives him to pursue his research, but rather because he is attempting to palliate what he (correctly) considered to be an antagonistic crowd. The quotation upon which Field bases his case, if viewed in its immediate context of Tarski's apprehensiveness, appears to be a rhetorical maneuver, instead of a manifestation of a deeply held and thoroughgoing philosophical orientation.

We cited Tarski's view that the semantic conception was philosophically neutral. Tarski's ambivalence toward philosophical theorising must also be recalled.[75] And though the question whether physicalism could underwrite the semantic conception is a valid question, the attribution of the view to Tarski seems to overlook key aspects of Tarski's philosophical approach, in view of the evidence.

It must be said that at the end of the day, labels (such as naturalism or even nominalism or physicalism) are useful here only up to a point. Tarski inherited the liberalism of the Polish School; in particular he inherited its experimental attitude toward foundations, the idea of speculation for its own sake, or for the sake of establishing theorems, and the idea that pursuing logical work should be done in the absence of ideological pressure.[76]

Hodges has written in [106] about this experimental attitude of the Polish School, in particular about the idea that deductive theories are something that one "practises" – indeed this is even encoded in the language, through the phrase *uprawiać sformalizowane nauki dedukcyjne*, meaning to "practise formalised deductive sciences":

> It does seem that in all his discussion of deductive theories Tarski has in mind a highly idealized mathematician. This mathematician is able to do mathematics by 'practising' deductive theories, and can switch his or her cognitive faculties on or off at will.

---

[74] Frost-Arnold also poses the urgent question whether the relevant sentence was translated correctly from the Polish original.

[75] We cited Frits Staal above, a friend and Berkeley colleague of Tarski's, recalling Tarski expressing the thought "that he did not like to talk much about philosophy because it was not rigorous". [57], p. 318.

[76] Think of the wonderful expression "metaphysicsism"!

Here the 'practising' logician, the (albeit idealised) logician who *uprawiać sformalizowane nauki dedukcyjne*, "practises" deductive theories from within the territory of natural language. What is the nature of this practice, how to name it? One might call it on-again-off-again-ism: the localised, dynamic and transient use of metamathematical ideas in logico-mathematical practice. The methodology is inscribed into Tarski's theorems – if we want to read those theorems that way. Of interest to us here is the fact that contemporary model theorists would develop the methodology greatly in subsequent work (see Chapter 6).

The tension inherent in maintaining a belief in mathematics' groundedness, a belief that Tarski chiseled into his logical practice even as it was masked by a publicly announced scepticism toward that very groundedness, places Tarski in the midst of what many philosophers would judge to be philosophical wreckage. But we are interested less in motivation than in certain developments in exactly this Tarskian vein in contemporary model theory, developments which would lead to an avowedly syntax free approach to model classes, pursued for purely programmatic (or technical) reasons, rather than for any reasons Tarski may or may not have had.

Syntax freeness nevertheless remains, especially in the context of model theory, aspirational, for reasons we will go into below. Do we ever get there? The situation is plastic; and there is noise coming from the fact that the syntax/semantics distinction can be fairly unstable. Here is Tarski in discussion at the 1965 International Colloquium in the Philosophy of Science:

> Now I would answer that this use of non-elementary notions [i.e second order logic JK] is not necessary in formalization. I have often stressed that it is not necessary in formalizing (if by that one means normal first order formalization) to define what is a formula, what is a proof, what is a theorem. Formalization may consist in a series of autobiographical statements. Of course the statements must include such sentences as: "I recognize the following statement as true, as an axiom, and I don't want to analyze it" (and I write down the axiom). Then I say, "I shall also recognize as true any sentence b under the condition that I have previously recognized as true two sentences, one of which is a and the second of which is $a \supset b$." And this is enough. You see that in this formalization I do not actually use any such theoretical term as set; and I do not need a definition of a sentence or a formula. I have said quite clearly what I am going to do in developing arithmetic, and this procedure is quite satisfactory for various purposes.[77]

---

[77] The meeting was held at Bedford College, London. Tarski was responding to Kreisel's suggestion that second order logic may be useful in solving open problems in set theory. Quoted in [209], p. 253. Tarski's remarks were transcribed from the tape recording of the discussion in the Tarski papers in Bancroft Library.

The passage is a curious mixture of the formal with the informal with the autobiographical, with concepts such as theoremhood, truth and proof. The language is vivid and interesting, and Tarski seems to have dispensed with the standard meanings of certain logical categories – categories which are, after all, fairly porous to begin with.

5.3 Squeezing First Order Definability

We permit ourselves a short digression on the topic of squeezing arguments. Tarski's attempt to isolate an adequate definition of definability in DSRN and [249] is clearly very different in character from Gödel's (and from Post's). Instead of seeking an absolute notion of definability, adequate in the sense of capturing the informal or intuitive sense of the term, Tarski's adequacy claim seems to rest on something much more in the direction of a squeezing argument.

Squeezing arguments were introduced by Kreisel in his 1967 "Informal rigour and completeness proofs" [137], a landmark paper in the area of informal methodology. Since then they have been taken up by W. Dean [50], H. Field [67], V. Halbach, P. Smith [240] and others, also in our [129].[78] Squeezing arguments have the following form:

Consider an informally defined mathematical concept I. Formally define two concepts A and B such that falling under the concept of A is a sufficient condition for falling under the concept of I, and falling under the concept of I suffices for falling under the concept of B. Thus $A \subseteq I \subseteq B$, where the inclusions are understood as applying to the extensions of the concepts A, B, I.

Now suppose the formal notions A and B have the same extension. Then by the inclusions $A \subseteq I \subseteq B$ the extension of the informal concept I must coincide with that of A and B.

For the informal concept I Kreisel took intuitive validity, denoted Val, understood as truth in all possible structures. This includes set and class-sized structures, as well as, in principle at least, structures that have no set-theoretical definition.[79] Taking formal first order provability, denoted D_F, on the left, and

---

[78] An early use of the term "squeezing argument" occurs in the postscript to Field's 1984 paper "Is mathematical knowledge just logical knowledge?", which appeared in Field's 1989[66].

[79] W. Dean has commented thus on this aspect of the squeezing argument:

In fact much of the context of 1967b is provided by Kreisel's engagement with potential 'multifurcations' (his word) of the notion of set, including (proper) classes, properties understood intensionally, and presumably also sets in forcing extensions. So one of his

taking truth in all set-theoretical structures,[80] denoted V, on the right, Kreisel argued as follows: By soundness, $D_F \subseteq$ Val. By the fact that truth in all structures entails truth in all set-theoretical structures, Val $\subseteq$ V. Thus

$$D_F \subseteq \text{Val} \subseteq V.$$

Invoking the completeness theorem for first order logic Kreisel concludes the following *theorem*, as he calls it, for α any first order statement:

$$\text{Val}\, \alpha \leftrightarrow V\alpha \text{ and Val}\, \alpha \leftrightarrow D_F \alpha.$$

Kreisel's argument does not depend on a proof of soundness in classical set theory, an issue Field takes Kreisel to task for in his [67]. Instead Kreisel takes soundness for granted, on the basis of *historical experience* – or as Kreisel puts it, *intuitive notions standing the test of time*. In Kreisel's terminology, Field would take D_F as "primary":

> First (e.g. Bourbaki) 'ultimately' inference is nothing else but following formal rules, in other words D is primary (though now D must not be regarded as defined set-theoretically, but combinatorially). This is a specially peculiar idea, because 99 per cent of the readers, and 90 per cent of the writers of Bourbaki, don't have the rules in their heads at all![81]

As for Kreisel's own "proof" of soundness, extending to α^i (interpreting α as an ith order sentence) for all i, he simply argues that the universal recognition of the validity of Frege's rules (D_F) at the time, together with the "facts of actual intellectual experience" accumulated subsequently, should amount to no more and no less than the acceptance of

$$\forall i \forall \alpha (D_F \alpha^i \to \text{Val}\, \alpha^i)$$

for us.[82]

Reading Tarski's DSRN as proposing a squeezing argument, *but for the intuitive notion of first order definability*, the argument would go as follows: Let SF$_D$ denote the metamathematical definition of definability via sentential

---

> underlying points seems to be that although relevant notion of "structure" in which intuitive validity is grounded historically may not select between these options, the squeezing argument should still convince us that Val = V.

> Dean, personal communication.

[80] A set-theoretical structure being one whose domain, relations and functions are sets in the usual sense.

[81] [137], p. 153.

[82] The above presentation of Kreisel's squeezing arguments is adapted from our [129].

functions, I_D denote intuitive definability, and B_D denote the mathematical definition of D_F via the Boolean operations and projection.

$$SF_D \subseteq I_D \subseteq B_D.$$

Invoking the mathematically established equivalence between SF_D and B_D of DSRN, we conclude the following *theorem*, for α any first order sentential function:

$$SF_D \leftrightarrow I_D \text{ and } I_D \leftrightarrow B_D.$$

Of this variant of Kreisel's squeezing argument W. Dean has asked the very important question about the pre-theoretical status of first-order definability, in particular whether there is "additional evidence that first-order definability really is a 'common', 'traditional', or 'intuitive notion' in Kreisel's sense – e.g. is it embedded historically in our practices in anything like the way validity or effective computability are?"[83] Our view is that such an argument can be given, though we will not give it here.

At the end of the day it is perhaps not surprising that the methodology of DSRN should slot so easily into the framework of Kreisel's squeezing arguments – both Kreisel and Tarski were beguiled by "the mathematical" in one form or another.

5.4 Tarski and Logicality

In the opening of his posthumously published lecture [252] called "What are logical notions?" Tarski remarks, wittily, that "specialists working in a given science are usually the people least qualified to give a good definition of the science". Tarski nevertheless goes on to propose a definition of the concept of logicality, or "logical notion". A caveat is given at the outset to the effect that he is not interested in pinning down the core meaning of the concept of logicality:

> Answers to the question 'What is logic?' or 'What is such and such science?' may
> be of very different kinds. In some cases we may give an account of the prevailing
> usage of the name of the science ... In other cases we may be interested in the
> prevailing usage, not of all people who use a given term, but only of people who
> are qualified to use it – who are expert in the domain. Here we would be interested
> in what psychologists understand by the term 'psychology'. In still other cases our
> answer has a normative character: we make a suggestion that the term be used in a

---

[83] Dean, personal communication.

certain way, independent of the way in which it is actually used. Some further answers seem to aim at something very different, but it is very difficult for me to say what it is; people speak of catching the proper, true meaning of a notion, something independent of actual usage, and independent of any normative proposals, something like the platonic idea behind the notion. This last approach is so foreign and strange to me that I shall simply ignore it, for I cannot say anything intelligent on such matters.

Tarski then proposes a "possible" use of the term "logical notion", based on the Erlanger programme due to Felix Klein, designed for classifying geo-metrical notions. What was the Erlanger programme? Here one declares the "notions" of metric geometry, descriptive geometry and topological geometry, to be those invariant under the respective transformations: similarity, affine, and continuous. Thus a topological notion, for example, will be one invariant under continuous transformations of the underlying topological space.

Tarski observes that for a given subject area, the concepts classified as invariant are inversely related to the number of transformations – the more transformations there are, the fewer invariant notions there are. If one then thinks of logic as the most general of all the mathematical sciences, why not declare "logical" notions to be the limiting cases? Thus a notion is to be thought of as *logical* if it is invariant under *all* permutations of the relevant domain.

Tarski notes that under this definition the $\in$ relation will not be construed as logical, taking ZFC set theory as the base theory, for the obvious reason that the relation is not invariant under the Tarski criterion. On the other hand, taking *Principia Mathematica* as the base theory, the $\in$ relation *will* be con-strued as logical. This is due to the equally obvious reason that the $\in$ relation does not appear as part of the syntax, but is rather simply built into the type structure. Thus any permutation of the underlying domain (i.e. the urelements), will leave the $\in$ relation unchanged.[84] In the terminology of this book, Tarski is detecting the entanglement of the $\in$ relation with two different formalisms: ZFC set theory on the one hand, and the type theory of the *Principia* on the other.[85]

---

[84] See also Lindenbaum and Tarski's "On the limitation of the means of expression of deductive theories" [251, pp. 384–392] in which it is shown that every relation definable in the simple theory of types is invariant under every permutation of the base domain, i.e. the domain of individuals.

[85] This is because a change in the logical framework induces a change in the formal environment, in particular the two systems differ in the judgement whether the $\in$ relation is a logical relation: the former system negatively, the latter system positively.

Tarski's suggestion has generated a substantial literature, with numerous arguments presented both for and against what is now known as the Tarski–Sher criterion.[86] Most of these arguments are characterised by the attempt to pin down the core notion (of logicality), so completely ignoring Tarski's caveat.

S. Feferman's [62], based on his earlier [60], in which he proposes a notion of *homomorphism* invariance, lays out a number of criticisms of the Tarski–Sher thesis:

> I critiqued the Tarski-Sher thesis in [60] on three grounds, the first of which is that it assimilates logic to mathematics, the second that the notions involved are not set-theoretically robust, i.e. not absolute, and the third that no natural explanation is given by the thesis of what constitutes the same logical operation over arbitrary basic domains.[87]

Feferman's first objection, namely the fact that Tarski sought to express the concept "logical notion" in mathematical terms, overlooks Tarski's lifelong programme to carry out exactly what Feferman accuses Tarski of here, namely assimilating logic to mathematics by expressing metamathematical concepts in mathematical terms – the very programme we dwell on at length in this chapter. Feferman's second, more pressing objection is that under the Tarski invariance criterion, a canonically mathematical but non-absolute notion such as cardinality becomes a logical notion, being invariant under permutation of the relevant underlying domain (or domains, if one has the Tarski–Sher criterion in mind). Tarski is of course aware that cardinality is construed as logical under his criterion, and indeed this is not a problem for him at all, but completely in keeping with the spirit of his logical and metamathematical work:

> That a class consists of three elements, or four elements ... that it is finite, or infinite – these are logical notions, and are essentially the only logical notions on this level.[88]

Feferman proposes restricting the Tarski–Sher invariance criterion to operations that are absolute with respect to set theories making no assumptions about the size of the given domain—a strong absoluteness assumption. Feferman cites [259],[89] in which it is shown that operations on relational structures

---

[86] G. Sher [230] generalised Tarski's invariance criterion to cover notions invariant across isomorphic structures.

[87] [62], p. 1.

[88] I.e. the level "classes of classes." [252], p. 151.

[89] Feferman also cites an unpublished result of K. Manders.

that are definable in an absolute way relative to KPU-Inf, i.e. Kripke–Platek set theory with urelements and without the Axiom of Infinity, are exactly those expressible in the ordinary first-order predicate calculus with equality. As an aside, and given the basic theme of this book, the result is interesting because it gives yet another characterisation of first order logic – though unlike Lindström's, it is not a semantic characterisation but a set-theoretical one.

D. Bonnay [24], G. Sagi [213] and others have followed the Feferman view, that logicality should be neutral also with respect to *cardinality*. Sagi's view in particular is that logicality should be thought of as a graded notion, graded by cardinalities in the form of Löwenheim numbers.[90] As Sagi writes in her [213]:

> I ... propose that among the isomorphism-invariant terms (where, loosely speaking, only cardinality is a factor), a term is more logical the less it distinguishes between different cardinalities.

Sagi proposes a criterion based on terms being "fixed faithfully" in standard model-theoretic semantics.[91] Under certain natural assumptions, terms that are faithfully fixed turn out to be isomorphism invariant. Sagi then turns to specific logical constants, namely the quantifiers Q of a logic $\mathcal{L}(Q)$, and proposes grading the logicality of these terms by their Löwenheim number.[92] The strategy here is to "view Löwenheim numbers as telling us how much structure a term requires in order to be fixed in the context of a logic". As Sagi observes,

---

[90] The Löwenheim number of a logic L, denoted here by $\ell(L)$, is the least cardinal μ such that any L-sentence in a countable vocabulary which has a model, has a model of cardinality less or equal to μ. Otherwise, $\ell(L) = \infty$.

[91] This means the following:

> We shall thus say that a term is fixed in standard model-theoretic semantics if its extension is a function of the domain, namely, it is constant across models sharing the same domain ... We shall thus say that a term associated with an intension is fixed in a manner faithful to its intended meaning, for short: fixed faithfully if it is fixed, and the value of the associated operation in every model represents the value of the intension in every possible world represented by the model (on some correspondence), and no other extension. In other words, for a term to be faithfully fixed, its extension in each model must not represent anything but its extension in some possible world.

[213], p. 9.

[92] $\mathcal{L}(Q)$ denotes first order logic with the generalised quantifier Q appended. By the Löwenheim–Skolem Theorem, the Löwenheim number of first order logic is $\aleph_0$; for each α, the Löwenheim number of the logic $\mathcal{L}(Q_\alpha)$ is $\aleph_\alpha$. The expression "$Q_\alpha x \phi(x)$" means that "there are at least $\aleph_\alpha$ many x such that $\phi(x)$".

... for a logic L, if $\ell(L) = \kappa$, then dismissing all models of cardinality greater than κ will not make a difference to the validities of the logic. ... We may venture to say that the terms in the logic L in such a case, if fixed faithfully, require no more structure than that which is given by the set of cardinalities less or equal to κ.

For example, the quantifier Q_α associated with the logic $\mathcal{L}(Q_\alpha)$, which, again, has Löwenheim number $\aleph_\alpha$, is indifferent to cardinalities greater than $\aleph_\alpha$; to the extent that the quantifier "has structure", this is determined only by the set of cardinalities less than or equal to $\aleph_\alpha$. Assuming that, as Sagi puts it, "the higher set-theoretic infinite is metaphysically loaded... The lower the cardinalities to which the meaning of a term may be sensitive, the more logical it is."

Sagi's parametric treatment of logics defines a calculus for measuring the *metaphysical* entanglements of a quantifier, that is, its entanglement with the set theoretic hierarchy construed as a scale of metaphysical commitments. Under this criterion, first order logic and $\mathcal{L}(Q_0)$ are maximally logical; and the grade of logicality decreases as α increases: if $\alpha \leq \beta$, then Q_α is more logical than Q_β. As for other quantifiers, the quantifiers "most", "more" and the equicardinality or Härtig quantifier all have the same (high) Löwenheim number, and thus qualify as less logical than Q_α.[93]

Sagi's classification of logicality cuts across entrenched partitions of logical space in interesting ways. According to Sagi, "We should distinguish between a logic $\mathcal{L}(Q)$ used to measure the logicality of Q and the logic we ultimately use for validity and logical consequence."[94] For example there is the issue of expressive power, with respect to which the logics built on the quantifiers "most", "more" and the equicardinality or Härtig quantifier differ.[95] But there is also the issue that some of these logics are axiomatisable and some are not: Keisler's axioms are complete for $\mathcal{L}(Q_1)$, and if GCH is assumed, then they are complete for all $\aleph_{\alpha+1}$ for which $\aleph_\alpha$ is regular (hence for $\aleph_n$ for all $n > 0$). Whereas while $\mathcal{L}(Q_0)$ satisfies the Keisler Axioms, these are not a complete axiomatisation of $\mathcal{L}(Q_0)$.[96]

Väänänen has suggested that completeness, in the form of having a complete axiomatisation, should be a measure of logicality. With the exception of Q_0,

---

[93] If ℓ_I is the Löwenheim number of the Härtig quantifier, then ℓ_I is always bigger than the first fixed point of the $\aleph$-hierarchy [76]. If $V = L$, then ℓ_I is bigger than the first inaccessible (if any exist) [262]. If $V = L^\mu$, then ℓ_I is bigger than the first measurable cardinal [260]. If Con("there is a super compact cardinal"), then Con($\ell_I <$ the first weakly inaccessible) [162]. If Con(ZFC), then Con($\ell_I < 2^\omega$) [262].

[94] ibid, p. 22.

[95] See [99].

[96] See [180]. See also Section 6.3.1.

one then would think of "there are very (i.e. uncountably) many" as a logical constant, the meaning of which is dictated by Keisler's axioms. From this point of view the quantifier Q_1 would be graded as having a higher degree of logicality than Q_0, as $\mathcal{L}(Q_1)$ is complete with respect to the Keisler Axioms, whereas $\mathcal{L}(Q_0)$ is not.[97]

For singular strong limit $\aleph_\alpha$ Keisler [119] proved a Completeness Theorem, but with different axioms. Thus there are (at least) two different "logical" concepts of "very many", one for $\aleph_\alpha$ the successor of regular and one for the singular case. (Successor of singular is open.) The two concepts have different logical content: Keisler's Axioms for Q_α are valid for all infinite α, and in some cardinalities (successor of regular) they have a Completeness Theorem, modulo the GCH, as was noted above. In some other cardinalities (singular) new axioms have to be added in order to get a Completeness Theorem. In the base case $\alpha = 0$ no effectively given additional axioms can be added to give a completeness theorem.

The key observation here is that logicality under this latter view is identified with the elimination of semantic content, this being delivered by the relevant completeness theorem. To put it another way, the completeness theorem enables the conversion of semantic content into syntactic content.

The idea of tying logicality to syntax via a completeness theorem is spelled out in some detail in Bonnay's recent "Carnap's criterion of logicality" [25]. Bonnay proposes a modification the programme Carnap laid out in the *Aufbau* [33], calling for logical expressions to be defined syntactically. The claim is that the modification can be defended against the well-known objections of Quine [203]. Here the absoluteness of a logic plays a crucial role in guaranteeing the *robustness* of the syntactic definition. Thus $L(Q_0)$ is an absolute logic because it has a recursive syntax, just like first order logic, and its semantics is absolute in transitive models of (even weak) set theory.[98] On the other hand, $L(Q_1)$ is not absolute, even though it has a recursive syntax.[99]

Burgess's [29] smooths out this small morass of irregularities somewhat, by forging a link between absoluteness and the idea of having a proof procedure (for strong languages). As he writes: "There is a quasi-constructive complete

---

[97] According to Väänänen a possible objection might come from the fact that Keisler's axioms are satisfied by many different quantifiers. Väänänen, personal communication.

[98] This is essentially because finiteness is absolute, i.e. Δ_1-definable. See Barwise "Admissible sets and Structures" [14], p. 38.

[99] This is due to the well-known fact that countability is not absolute: a set can be uncountable in a model of set theory and countable in a transitive extension. On the relevance of absoluteness in this context see also Bonnay's [24].

proof procedure involving rules with $\aleph_1$ premisses for any strong first order language ..."[100]

Given these two frameworks grading logicality, or indeed any other such framework, one may ask, if one classifies the first order existential quantifier "there is at least one", as *inherently logical*, and if as such this quantifier is thought of as having minimal or no semantic content, then is there a principled way to determine when and how higher quantification acquires semantic content, e.g. at what level in the cumulative hierarchy? Sagi's analysis provides one answer: logicality diminishes the higher up we are in cumulative hierarchy; while Väänänen suggests that logicality kicks in arbitrarily high up, e.g. for all $\aleph_{\alpha+1}$ for which $\aleph_\alpha$ is regular, with the $\aleph_0$ case an anomaly, again because of completeness.

As for the philosophical consequences, Sagi's graded account of logicality pays heed to, indeed incorporates, the standard view of set theory, namely that it is metaphysically involved. In particular one's metaphysical commitments – which *are* semantic commitments, under the standard view – increase as one moves up the cumulative hierarchy. While one would expect that Bonnay's account (and defence) of Carnap's treatment of logical expressions is at the same time a defence of syntax-based conventionalism (in Carnap's formulation).

Logicality encodes a form of indifferentism, indifference to subject matter, while formalism freeness encodes the idea of indifference to the underlying logic or formalism. In that sense logicality may be thought of as formalism freeness's logical cousin. Formalism freeness as we defined it "skews semantic", so under the Väänänen view, logicality would lie at the opposite end of the spectrum from formalism freeness, while under the Sagi view logicality is semantically involved, and in that sense closer to Tarski's account of logicality.[101]

Of course formalism freeness is built into the Carnapian framework, if it is not even its bedrock principle, through the Principle of Tolerance. As Bonnay puts it:

> according to the principle of tolerance, there is a great variety of possible language-forms which have equal rights to be considered as the basis for the logic of science. To account for these other possible frameworks, Carnap outlines in the fourth part of the *Logical Syntax of Language* [33] a general theory of syntax.

---

[100] Burgess defines a strong first order language to be an infinitary language "which is strong enough to express well-foundedness, at least over countable structures, yet weak enough that the satisfaction relation is Δ_1".

[101] Sagi's [213], to which the reader is referred, has been summarised all too briefly here.

General syntax aims at characterizing the key concepts of syntax independently of the choice of a particular language.[102]

For the programme of the *Aufbau* to succeed it was necessary to define general syntax in a *formalism free* way. In his review [157] of the English translation of the *Aufbau* S. Maclane calls attention to the difficulty of carrying this out:

> The points discussed above show how difficult is the task of defining so many relatively specific concepts in an absolutely arbitrary language. The notion of "any language" may be just as treacherous as was the notion of "any curve" before the critique of analysis situs.[103]

To treat Carnap's notion of general syntax in the depth required would take us too far afield. What is of interest to us is the fact that Carnap stands with Tarski and with Gödel in asking for a language transcendent formulation of a key metamathematical concept. In Carnap's case, the concept of a formal language itself; in Tarski's case, a syntax-free notion of the concept "logical constant".

5.5 In Sum: Parataxis

The ease with which Tarski carried out his programme to formulate meta-mathematical theorems in semantic terms – terms, that is, that mathematicians would be able to understand – masks the radical naturalism at the heart of his natural language project. The fact that Tarski's philosophical reticence was nearly complete, taken together with Tarski's overtly expressed sympathy with nominalism, or at times physicalism, or at times even finitism, complicates the interpretive task at hand. The term *parataxis* comes to mind, a device in poetry characterised by juxtaposition of distinct textual elements, with no attempt at linkage – and yet a paratactic passage may have an underlying discernible meaning. Here is an example of parataxis from Ezra Pound's Canto LXXXI:

> sky's clear
> night's sea
> green of the mountain pool

---

[102] Bonnay, [24] Of the vast literature on logicality the reader is also referred to Bonnay and Westerstahl's [26], solving Carnap's problem (in [34]) of how to rule out non-standard interpretations of the logical constants which nevertheless validate all the classical laws.
[103] MacLane, ibid. "Analysis Situs" refers to Poincaré's 1895 article [194].

How to resolve Tarski's paratactic practice? Tarski's project was pursued within a specific philosophical milieu, one for which the problem of accounting for apparently referring terms in mathematical discourse created a philosophical crisis. The crisis locked philosophers into a problem, that of giving a uniform semantics covering reference in both the empirical and mathematical cases.

In the attempt of knowledge cultures to free themselves from being tethered to a problem, the idea of *cultural anticipation* becomes relevant: the seemingly forward knowledge of subterranean movements of culture, when the overt markers of such tectonic movement are not yet in view. Such anticipatory events can only be marked by parataxis – in Tarski's case of the kind, that what is done comes apart from what is said.

We have suggested that Tarski's parataxis may mask an underlying connected field. We will leave this interpretive task for others. The attempt here was to fold into Tarski's philosophical world picture, what it was, in metamathematics, that he actually did.

5.6 Coda: An Improvement of McGee's Theorem

An oft-cited theorem on logicality is due to V. McGee [175] and states that logical notions are definable in $L_{\infty\infty}$ for each cardinality separately. That is:

Suppose $\mathcal{K}$ is a class of models of a fixed vocabulary L. Without loss of generality we assume $L = \{R\}$, where R is a binary predicate symbol. For any cardinal λ let $\mathcal{K}_\lambda$ be the class of elements of $\mathcal{K}$ with a domain of size λ. McGee [175] proved that the following are equivalent:

(1) $\mathcal{K}$ is closed under isomorphisms.
(2) For every λ, the class $\mathcal{K}_\lambda$ is definable in $L_{\infty\infty}$.

It is important to note that the definition depends on the cardinality of the models in the class. We will present another way of defining a model class in the next chapter, and in the last chapter we will compare the two methods. For now we ask the question, is McGee's theorem optimal? Väänänen has proved that (1) above is equivalent to:

(3) For every λ, the model class $\mathcal{K}_\lambda$ is definable in $\Delta(L_{\infty\omega})$.[104]

---

[104] Väänänen, unpublished manuscript. For the definition of the Δ operation see Section 6.6.

Why is Väänänen's result an improvement? First, $\Delta(L_{\infty\omega})$ is a proper sublogic of $L_{\infty\infty}$. $\Delta(L_{\infty\omega})$ also has a strong Löwenheim–Skolem Theorem: If a sentence of $\Delta(L_{\kappa^+\omega})$, κ regular, has a model of cardinality $\lambda > \kappa$, it has models of all cardinalities μ with $\kappa \leq \mu \leq \lambda$. This is in contrast to $L_{\infty\infty}$, which does not have a Löwenheim–Skolem Theorem in the same strong form as $\Delta(L_{\infty\omega})$. Secondly, the logic $L_{\infty\omega}$ is an absolute logic and $\Delta(L_{\infty\omega})$ inherits much of the absoluteness of $L_{\infty\omega}$, while $L_{\infty\infty}$ is badly non-absolute. For example, the sentence of $L_{\infty\infty}$ which has models exactly in cardinalities μ such that $\mu^\omega = \mu$ is highly non-absolute. Finally, the Hanf number of $\Delta(L_{\kappa^+\omega})$ is only moderately large, namely $< \beth_{(2^\kappa)^+}$, while the Hanf number of $L_{\omega_1\omega_1}$ is bigger than the first weakly inaccessible cardinal (and consistently bigger than the first measurable cardinal).

To prove that (3) above implies (1), we first introduce some formulas due to D. Scott [215]. For any ordinal α let the formula $\eta_\alpha(x)$ with one free variable x and a binary relation symbol $<$ be defined, by transfinite recursion, as follows:

$$\eta_\alpha(x) \leftrightarrow \forall y(y < x \rightarrow \bigvee_{\beta<\alpha} \eta_\beta(y)) \wedge \bigwedge_{\beta<\alpha} \exists y(y < x \wedge \eta_\beta(y)).$$

Then for a linear order $(A, <)$ we have

$$(A, <) \models \eta_\alpha(a) \iff (\{b \in A : b < a\}, <) \cong (\alpha, <).$$

Let

$$\eta'_\alpha \leftrightarrow \forall y \bigvee_{\beta<\alpha} \eta_\beta(y) \wedge \bigwedge_{\beta<\alpha} \exists y \, \eta_\beta(y).$$

Now for a linear order $(A, <)$ we have

$$(A, <) \models \eta'_\alpha \iff (A, <) \cong (\alpha, <).$$

Suppose $\mathfrak{A}$ is a model of the vocabulary L and $|A| = \lambda$. Let $f_A : \lambda \to A$ be a bijection and $a <_A b \iff f_A^{-1}(a) < f_A^{-1}(b)$. For $\alpha, \beta < \lambda$ let

$$\rho_{\alpha,\beta}(x, y) = \begin{cases} R(x, y) & \text{if } \mathfrak{A} \models R(f_A(\alpha), f_B(\beta)) \\ \neg R(x, y) & \text{if } \mathfrak{A} \not\models R(f_A(\alpha), f_B(\beta)). \end{cases}$$

Let

$$\Phi_\mathfrak{A} \leftrightarrow \forall x \forall y \bigwedge_{\alpha,\beta<\lambda} ((\eta_\alpha(x) \wedge \eta_\beta(y)) \rightarrow \rho_{\alpha,\beta}(x, y)).$$

Now $(\mathfrak{A}, <_A) \models \eta'_\lambda$ and $(\mathfrak{A}, <_A) \models \eta_\alpha(a) \iff a = f_A(\alpha)$. Hence

$$(\mathfrak{A}, <) \models \eta'_\lambda \wedge \Phi_\mathfrak{A}.$$

On the other hand,

$$(\mathfrak{A}', <') \models \eta'_\lambda \wedge \Phi_\mathfrak{A} \ \Rightarrow \ \mathfrak{A}' \cong \mathfrak{A},$$

for if $(\mathfrak{A}', <') \models \eta'_\lambda \wedge \Phi_\mathfrak{A}$ and $g : (A', <') \cong (A, <_A)$, then $g : \mathfrak{A}' \cong \mathfrak{A}$. Let

$$\Theta_\mathcal{K} \leftrightarrow \bigvee \{\Phi_\mathfrak{A} \ : \ \mathfrak{A} \in \mathcal{K}, A = \lambda\}.$$

Now by the above,

$$\mathfrak{A} \in \mathcal{K}_\lambda \quad \Longleftrightarrow \quad (\mathfrak{A}, <) \models \eta'_\lambda \wedge \Theta_\mathcal{K} \text{ for some } <$$
$$\Longleftrightarrow \quad (\mathfrak{A}, <) \models \eta'_\lambda \rightarrow \Theta_\mathcal{K} \text{ for all } <.$$

Since $\eta'_\lambda, \Theta_\mathcal{K} \in L_{(2^\lambda)^+ \omega}$, we are done.

6

Model-Theoretic Aspects

Now, to the second sailing of model theory! Let us frame it now by considering again two important old concepts (of Patochka): This is about the contrast between describing, telling explicitly axiomatizing a theory ... on one hand, and looking at how different variants of a structure 'fit' within one another, how they 'reflect' in the small properties of the large, how they form a coherent description provided to us without obvious reference to phrases, sentences, formulas, theories.[1]

When we shift from the domain of mathematical logic to the more restricted area of research in model theory, we notice a continuation of the movement away from the leading role of formal languages in certain key areas. We have already met with the Tarskian idea that definable sets of reals (just as the definable sets of any o-minimal structure) can be viewed in a natural way as elements in a sequence of Boolean algebras closed under projection mappings and other mathematical properties, an entirely logic free notion of definability.[2] Moving forward there is the work on model classes of Jónsson and Fraïssé, who looked at model classes the way Tarski did, that is to say *mathematically*. Thus Fraïssé [72] presented conditions (joint embedding property, closure under substructures, amalgamation) on a class of finite models which are both sufficient and necessary for the class being the class of finite submodels of a unique countable model, called the Fraïssé limit of the class. And Jónsson [114] generalised the Fraïssé construction to higher cardinalities.[3] As for contemporary model theorists Shelah and others were

---

[1] A. Villaveces, lecture "Grasping Smoothly and Letting Go" Helsinki Collegium for Advanced Studies, 2015. Villaveces is referring to Jan Patočka's "La genèse de la rèflexion européenne sur le Beau dans la Gréce Antique," in the collection of essays *L'art et le temps*, P.O.L. 1990 (translated from Czech to French by Erika Abrams).

[2] See also Shelah's [226]. See also Pillay and Steinhorn's [193].

[3] Jónsson's construction is used in AECs to show that the class contains a so-called "monster model", see Section 6.1.

motivated in relevant instances toward syntax freeness by the desire to find a general and natural framework for results such as Morley's Theorem, a cornerstone of first order model theory,[4] or the desire to work in a framework without compactness.[5] The term "pure semantic method," due to Baldwin [9], seems applicable; as is the following remark of Saharon Shelah, describing the thesis of his [226, 227]: "Considering classical model theory as a tower, the lower floors disappear – compactness, formulas, etc.... the higher floors do not have formulas or anything syntactical at all."[6]

Model theory is a vast and flourishing mathematical subject, and it lies beyond the scope of this book to give anything like a comprehensive account of how the syntax/semantics distinction's shifting sands have transformed model theory into the field it is today. We nevertheless take note of certain developments relevant to the theme of this book, developments which come both from the first order and from the higher order/infinitary logic side of the practice.

In fact Tarski's natural language orientation, his "trust" in natural language, as we put it, both did and did not lead to a kind of pure semantic practice – though many would predict that things are moving in that direction, with first order model theory lately developing into a branch of analysis (both complex and real) and algebraic geometry.[7] In certain stability theoretic contexts, for example in the Zariski structures of Zilber and Hrushovski [109], or the Tarski systems of van den Dries [270], there is a tendency to emphasise, over the underlying first order logic, the relevant semantic characterisation of the logical concept. What is important for the geometric or algebraic study of structures in stability theory are the closure and independence properties of definable sets and types, and they can, arguably, be simply listed without any reference to the syntax and semantics of first order logic.[8] What is relevant

---

[4] We are referring here to Shelah's [226, 227]. Morley's Theorem says in first order logic that a countable theory that is categorical in one uncountable cardinality is categorical in every uncountable cardinality.

[5] See [12] and [13].

[6] Shelah, personal communication.

[7] To quote Anand Pillay, "it is quite reasonable and informative to say that model theory is part of functional analysis". Pillay, "Remarks on purity of methods", to appear.

[8] As Villaveces puts it, "[also important] are various geometrical configurations, genericity of types, eventual behaviour of types (stationarity), domination of types, orthogonality of types ... 'the new primitives'. In FO Stability, of course, all their definitions are given syntactically, at least initially. However, many of these benefit from general, non-logical, formalism free, semantic characterisations. For instance, Adler ([2]) ... characterises forking and thorn-forking purely in terms of a 'geometry of possible extensions' of models and types (orbital types), ultimately giving up formulas (where the original definitions were apparently very formula-laden)." Villaveces, personal communication.

here are facts of the following kind: that quantifier elimination holds in a (one-dimensional) Zariski structure, is just the statement that the projection of a constructible set is constructible.

This development has transformed the model theorist's study of the "old" concrete objects, such as algebraic varieties and analytic curves, into one in which these are viewed in a new framework, one which generalises the classical approaches, but is spelled out with only passing reference to formal languages or their properties.[9]

Of interest to us as we track the profile of natural language in foundational practice is the fact that the blossoming of syntactic and semantic methods in the decades subsequent to Tarski's work pushed model theory even more in the direction of what we called *on-again-off-again-ism*, by which we mean the localised, dynamic and transient use of metamathematical ideas in logico-mathematical practice. This drew Hodges's attention to the Polish School, as we noted, who wrote of the mathematician being able to do mathematics by 'practising' deductive theories, that is, by switching his or her cognitive faculties on or off at will.[10]

6.1 Abstract Elementary Classes

One of the many episodes we could have chosen to illustrate the semantic approach in contemporary model theory is Shelah's notion of *abstract elementary class*, an entirely syntax free generalisation of Tarski's notion of elementary class.

Shelah defined the concept of an abstract elementary class in the 1980s, when trying to develop the model theory of infinitary languages and their extensions by generalised quantifiers. Shelah found that the syntax became an impediment, so he eventually decided to dispense (in this context) with the language altogether and merely state the properties he needed classes of models to satisfy, whether these properties arose from the syntax and semantics

---

[9] For a finely grained analysis of various formalism free phenomena in the stability context, the reader is referred to John Baldwin's [10], also to his earlier "Formalization, primitive concepts and purity" [9]. The reader is also referred to Hodges's "What is a structure theory?", in particular the discussion around the remark that "The difficulties about *aligning algebra with logic* haven't prevented Shelah from using the notion of superstability to prove results with a clear algebraic content." [104], p. 214, emphasis ours. For a general overview of the usefulness of first and higher order methodology in mathematics proper and as a supplement to Baldwin's [10] see Manin [170]. See also Robinson's retirement address [208], which contains a list of mathematical problems Robinson thought would be solvable using methods from logic and model theory.

[10] See Section 5.4.

of any particular infinitary language or not.[11] Model classes that satisfy these assumptions came to be known as AECs (for abstract elementary classes). A typical non-trivial assumption is that the model class is closed under unions of chains with respect to an abstract (strong) submodel relation, as one has with the elementary submodel relation.

In his introductory note to [229] Shelah discusses his motivations for introducing abstract elementary classes:

> A syntax-free way to treat model theory is to deal with Abstract Elementary Classes (aec). Why were they introduced ([Sh:88])? Looking at sentences in the logics $L_{\aleph_1,\aleph_0}$ or $L(Q)$ or their combinations, with exactly one or just few models in $\aleph_1$ (up to isomorphism, of course, [Sh:48]), there was no real reason to choose exactly those logics. We can expand the logic by various quantifiers stronger than Q (there are uncountably many), and still have similar results. The answer was to try to axiomatize "an elementary class" using only the most *basic*[12] properties. This, of course, does not cover some interesting cases (like the $\aleph_1$-saturated models of a first order theory, or classes of complete metric spaces), but still, it covers lots of ground on one hand, and has an interesting model theory on the other.

Tarski gave a characterisation of universally axiomatised theories in terms of certain closure properties of a class of structures. Recalling that an elementary class is the class of models of a first order theory, here is Shelah's definition of an *abstract* elementary class:

Definition 6.1.1 (Shelah) An Abstract Elementary Class (AEC) is a pair $\langle \mathcal{K}, \preceq_{\mathcal{K}} \rangle$ where $\mathcal{K}$ is a class of structures for some vocabulary L closed under isomorphism[13] and $\preceq_{\mathcal{K}}$ is a binary relation on $\mathcal{K}$ satisfying:

1. $\preceq_{\mathcal{K}}$ is a partial order.
2. For $M, N \in \mathcal{K}$, if $M \preceq_{\mathcal{K}} N$ then M is a substructure of N.
3. If $M_1 \preceq_{\mathcal{K}} N$ and $M_2 \preceq_{\mathcal{K}} N$ and $M_1 \subseteq M_2$, then $M_1 \preceq_{\mathcal{K}} M_2$.
4. $\mathcal{K}$ is closed under unions of chains and furthermore if each element of the chain is a strong submodel of some fixed $N \in \mathcal{K}$, then so is their union.
5. $\langle \mathcal{K}, \preceq_{\mathcal{K}} \rangle$ has a Löwenheim–Skolem number.[14]

For example: If $\mathcal{K}$ is the class of models of a first order theory T in a language $\mathcal{L}$ and $\preceq_{\mathcal{K}}$ is elementary substructure, then $\langle \mathcal{K}, \preceq_{\mathcal{K}} \rangle$ is an AEC.

---

[11] We use "language" here as meaning a logic, not a vocabulary.

[12] emphasis ours.

[13] i.e. $\mathcal{K}$ is a model class.

[14] I.e. there is an infinite cardinal κ, of size at least of the vocabulary of K, such that if A is any subset of the universe of $M \in K$, then there is $N \in K$ such that $N \preceq_{\mathcal{K}} M$, $A \subseteq N$ and $|N| \leq |A| + \kappa$.

This is of course the canonical example, but infinitary logics also give rise to AECs, such as $L_{\omega_1,\omega}(Q)$, as Shelah mentions above, or $L_{\kappa,\omega}$. Also Zilber's pseudo-exponential fields form an AEC (see below).

The five axioms for AEC's mimic the elementary substructure relation, but note that there is no mention of a formal theory or formal language. As Kueker describes them:

> Note that this definition is purely set-theoretic. In particular there is no syntax, and neither K nor $\preceq$ is assumed to be defined in any way. Although there are no formulas, Shelah gave a model-theoretic definition of types for AECs satisfying amalgamation (over models).[15]

and J. Baldwin describes Shelah's definition thus:

> Shelah's concept of an Abstract Elementary Class gives axiomatic but mathematical definitions of classes of structures in a[n informal] vocabulary $\preceq$. That is, the axioms are not properties expressed in some formal language based on $\preceq$ but are mathematical properties of the class of structures and some relations on it.[16]

The Joint Embedding Property (JEP), which holds in case of complete first order theories, is not assumed; similarly amalgamation, which holds in the presence of compactness, is now taken as an extra assumption. As for syntax freeness, in the abstract elementary class framework there is a sense in which a syntactic concept such as the concept of a formula (or more exactly of a "type") becomes, simply, a set invariant under automorphisms[17] – though strictly speaking, in the AEC framework there simply *is* no notion of formula.[18]

With the AECs Shelah in a sense isolated from infinitary languages the part that was susceptible to model-theoretic development. Subsequently the connection to infinitary languages has been forgotten and AECs are studied on their own.[19] In order to be able to isolate the axioms of AECs (such as the closure under unions of chains), it was essential to have the language in the background, but once this initial stage was reached, the language could

---

[15] [145].

[16] Baldwin, "Formalization, primitive concepts and purity," [9]. See also [10].

[17] Assuming the so-called Amalgamation Property, and arbitrarily large models in the class. This is the notion of a Galois type over a model, essentially an orbit under a group action – "an entirely mathematical concept". Baldwin, personal correspondence.

[18] The development of language freeness in this area of model theory within the Helsinki Logic Group was a primary stimulus for the writing of this book.

[19] See however [145], an interesting "return" to the roots of AECs in infinitary languages.

be dispensed with. In fact S. Vasey describes the AECs in even more general terms, as simply providing a semantic framework for model theory tout court.[20]

A typical "test question", in Shelah's argot, is to prove some form of Morley's Categoricity Theorem. Thus Shelah has conjectured that there is a cardinal κ such that if an AEC is categorical in some cardinal greater than κ then it is categorical for all $\mu > \kappa$.[21]

The AEC's have what Baldwin has called an "implicit" syntax,[22] i.e. any AEC can be defined in a suitable infinitary language:

Theorem 6.1.2 (Shelah) *For any AEC K in vocabulary L, there is a vocabulary $L' \supseteq L$, a first order theory T' in L', and a set of T'-types Γ such that if K' is the class of L'-models of T' that omit Γ, then $K' \restriction L = K$. Hence K is definable in $L_{\kappa^+ \omega}$ for a suitable κ and with extra predicates.*[23]

Hanf proved the infinitary languages have a Hanf number,[24] ergo via the Presentation Theorem the AECs obtain the same upper bound for the Hanf number ($< \beth_{(2^\kappa)^+}$). Thus Shelah obtains a semantic result by using the implicit syntax as a bridge. As Baldwin puts it:

> passing through the syntax, Shelah obtains a purely semantic theorem ... The syntactic condition in the theorem is a set of sentences in roughly Tarski's sense ... but we are able to deduce purely semantical conclusions.[25]

Baldwin points out an interesting further detail of the construction, namely that the logic giving the definition of the class is not connected to the class in any way otherwise; that is to say, there is no question of finding an *adequate* description of the class:

> via the presentation theorem, we are able to deduce purely semantical conclusions passing through the syntactic representation. Notably, the vocabulary arising in the presentation theorem arises naturally only as a tool to prove that theorem. *There is*

---

[20] Vasey, MathSciNet review of [11].

[21] This is Shelah's "Eventual Categoricity Conjecture". An AEC is categorical in κ if all models of cardinality κ that are in the class are isomorphic.

[22] [10], p. 298, ft. 19.

[23] [224].

[24] The Hanf number of a logic is the least κ such that if a sentence of the logic has a model of cardinality at least κ, it has arbitrarily large models. Respectively, the Hanf number of an AEC is the least κ such that if the class contains a model of cardinality at least κ, it contains models of arbitrarily large cardinality.

[25] [10], p. 300. See also [9], p. 10.

no apparent connection of each symbol of the resulting vocabulary with any basic mathematical properties of the AEC in question.[26]

The formal language $L_{\kappa^+\omega}$ serves then as a waystation, a stop along the way toward proving a theorem stated in a formalism free manner.[27] Hilbert emphasised the idea of an uninterpreted formalism; here the logician takes a step further, widening the gap between a (potentially adequate) formal vocabulary and the target structure.

J. Baldwin and W. Boney discuss the reintroduction of syntactic arguments in the context of Shelah's Presentation Theorem in their [11]:

> The interplay between syntax and semantics is usually considered the hallmark of model theory. At first sight, Shelah's notion of abstract elementary class shatters that icon. As in the beginnings of the modern theory of structures [43], Shelah studies certain classes of models and relations among them, providing an axiomatization in the Bourbaki [27] as opposed to the Gödel or Tarski sense: mathematical requirements, not sentences in a formal language. This formalism-free approach [124] was designed to circumvent confusion arising from the syntactical schemes of infinitary logic; if a logic is closed under infinite conjunctions, what is the sense of studying types? However, Shelah's Presentation Theorem and more strongly Boney's use [23] of AEC's as theories of $L_{\kappa^+\omega}$ (for κ strongly compact) reintroduce syntactical arguments.

The key word in the phrase "implicit syntax" is, of course, the word "implicit". What makes the syntax or logic of a target area implicit to the area? Shelah's Presentation Theorem makes free use of Skolem functions; Baldwin and Boney obtain a weaker result but replacing Skolem functions with Skolem relations in their 2017 [11], repairing a certain lack of canonicity in the original formulation of the theorem. As Baldwin and Boney describe their result:

> The requirement of disjointness in the syntactic arguments stems from a lack of canonicity in Shelah's Presentation Theorem: a single model has many expansions which means that the transfer of structural properties between an AEC **K** and its expansion can break down. To fix this problem, we developed a new presentation theorem, called the relational presentation theorem because the expansion consists of relations rather than the Skolem-like functions from Shelah's Presentation Theorem.[28]

---

[26] [10], p. 300. Emphasis ours.
[27] Baldwin, [9].
[28] [11], p. 330.

Whereas Shelah, and then Baldwin and Boney proved that any AEC can be viewed as a PC-class in an infinitary logic, Villaveces and Shelah[29] recently proved that any AEC can be viewed as an EC-class in an infinitary logic. Recalling that an EC class relative to a logic $\mathcal{L}^*$ is simply the class of models of an $\mathcal{L}^*$-sentence, a PC-class is a reduct of an EC-class:

Definition 6.1.3 A model class K is PC in the logic $\mathcal{L}^*$ if there is $\phi \in \mathcal{L}^*$ such that for all M

$$M \in K \leftrightarrow \exists \vec{R}[(M, \vec{R}) \models \phi].$$

Thus in the result of Villaveces and Shelah the extra predicates are not needed, in fact the class is definable in L_κ^1.[30]

6.2 Patchwork Foundations, On-Again-Off-Again-Ism and Implicit Syntax

It might be thought that eliminating the arbitrary aspects of the proof of the Presentation Theorem, arbitrary in the sense that extra predicates are needed to expand the language of the AEC, may open the door to re-establishing the adequacy of the logic. Adequacy aside, a complex of pragmatic and philosophical imperatives, e.g. the need to develop the model theory of infinitary and strong logics, and on the philosophical side the need to obtain categoricity theorems,[31] led to a kind of patchwork practice: syntactic concepts are introduced mid-proof, for pragmatic reasons, and then dropped to obtain a semantic theorem; and on the other hand a semantic framework is drawn on mid-proof, but for the sake of proving a syntactic theorem (see below for another example of both). Logical practice becomes localised, prismatic, and opportunistic, so that the model theorist is, at best, a local foundationalist: wanting to shed light on a certain area of the practice, but otherwise eschewing any attempt to supply a global foundation for it.

Baldwin [10] expresses the view very clearly:

> We approach global mathematical issues not by seeking a common foundation but by finding common themes and tools for various areas, not in terms of the topic

---

[29] F1632, unpublished manuscript.
[30] $\mathcal{L}_\kappa^1$ was defined in Section 2.2.3.
[31] We see the search for categorical axiomatisations of canonical theories as a fundamentally philosophical project, though we do not argue for it here.

studied, but *in terms of common combinatorial and geometrical features isolated by formalizations of each area.*[32]

But our interest is more in the local foundations of, say, plane geometry or differential fields. We set the stage for developing Thesis 1, by focusing on a specific vocabulary, designed for the topic rather than a global framework.[33]

As an aside, it has been suggested that the idea of "patchwork foundations" is already implicit in Tarski's work:

Methodology was not thought of as a single, unified science to be reduced ultimately to a single comprehensive deductive discipline. Moreover, Tarski does not indicate any single scheme for dividing it into separately codifiable subsciences. In fact, one aspect of Tarski's approach that adds clarity and flexibility is his willingness to isolate manageable subsciences according to the needs at hand and to aim at relatively comprehensive treatment of a limited subject matter instead of being forced to settle for a more limited treatment of a comprehensive subject matter.[34]

Baldwin devotes a substantial part of [10] to describing the details of what he identifies as a *paradigm shift* in model-theoretic practice, a progression from studying logics, to theories, to classes of theories (and their models):

The paradigm around 1950 concerned the study of logics; the principal results were completeness, compactness, interpolation and joint consistency theorems. Various semantic properties of theories were given syntactic characterizations but there was no notion of partitioning all theories by a family of properties. After the paradigm shift there is a systematic search for a finite set of syntactic conditions which divide first order theories into disjoint classes such that models of different theories in the same class have similar mathematical properties. In this framework one can compare different areas of mathematics by checking where theories formalizing them lie in the classification.[35]

The shift results, among other consequences, in the separation of first order model theory and axiomatic set theory:

---

[32] [10], p. 13.

[33] p. 29. Theses 1 and 2 are as follows: (1) Contemporary model theory makes formalisation of specific mathematical areas a powerful tool to investigate both mathematical problems and issues in the philosophy of mathematics (e.g. methodology, axiomatisation, purity, categoricity and completeness). (2) Contemporary model theory enables systematic comparison of local formalisations for distinct mathematical areas in order to organise and do mathematics, and to analyse mathematical practice. See [10], p. 29.

[34] J. Corcoran, editorial introduction to [251], xviii.

[35] [10], p. 2.

Here the paradigm shift enables the separation of first order model theory from axiomatic set theory.

... we explain this major aspect of the paradigm shift as the decision to choose definitions of model theoretic concepts that reduce the set theoretic overhead.[36]

The shift we identify here is enabled partly by the move to classify first order theories based on syntactic criteria. But where Baldwin's paradigm shift identifies the progression to a new set of goals within model theory, our shift toward increasing degrees of on-again-off-again-ism pays witness to a kind of final rupture with the idea of global foundations, the idea of logic used in any way foundationally in model-theoretic practice; and an even stronger merging of model theory with mathematics in natural language under the guise of "basic mathematical properties".

Let us look now at an example in which *semantics* serves as a waystation on the path to proving now a *syntactic* result. Zilber's notion of quasi-minimal excellent class is defined in [290]. Zilber's pseudo-exponential field $\langle F, +, \cdot, e^x \rangle$ is quasi-minimal excellent; it is therefore categorical in all uncountable powers, i.e. there is a unique model in every uncountable cardinality. Zilber's notion is fully semantic:

Zilber's notion of a quasi-minimal excellent class [290] was developed to provide a smooth framework for proving the categoricity in all uncountable powers of Zilber's pseudo-exponential field... The key point is that there are no axioms in the object language of the general quasiminimal excellence theorem; there are only statements about the combinatorial geometry determined by what are in the application the $L_{\omega_1 \omega}$-definable sets.[37]

But then Zilber proved in [290] that fields with pseudo-exponentiation are definable in $L_{\omega_1 \omega}(Q_1)$.[38] Thus whereas the Presentation Theorem is used to pass through the syntax for the sake of proving a semantic theorem (upper bounds for Hanf numbers), in a kind of converse, Zilber's result "passes through a (substantial) formalism free step in the argument to get a formal result – a categorical theory axiomatised in a formal language".[39] Zilber has conjectured that $\langle \mathbb{C}, +_{\mathbb{C}}, \cdot_{\mathbb{C}}, e_{\mathbb{C}}^x \rangle \cong \langle F, +_F, \cdot_F, e^x \rangle$, i.e. the pseudo-exponential field in power continuum is isomorphic to the complex field, the complex numbers with exponentiation.

---

[36] ibid, p. 314.

[37] Baldwin, "Formalization, primitive concepts and purity" [9].

[38] Q_1 is the generalized quantifier "there exists uncountably many". J. Kirby proved a more general result in [130]: if a quasi-minimal AEC has an infinite dimensional model, then the class is definable in $\mathcal{L}_{\omega_1, \omega}(Q_1)$.

[39] Baldwin, ibid.

Perhaps the most dramatic example of the opportunistic use of formal systems is the proof of the Mordell–Lang Conjecture for function fields due to E. Hrushovski – a purely mathematical conjecture with apparently no connection to logic, and at the same time a triumphant vindication of first order methods in model theory. Hrushovski formulated the conjecture in first order terms, and used stability theoretic methods to prove it [108]. As consequence, the purely mathematical result follows.

These are a few examples out of the many that could have been chosen, following this paradigm: foundational or logical concepts used pragmatically and as a tool, rather than in the service of foundational interests. The edifice of foundational theories becomes a patchwork edifice, and we see in many theorems a dynamic oscillation, moving from syntax to semantics and back again, as the pragmatics of the situation require.[40]

6.3 Implicit Syntax, Implicit Logic

The question of implicit syntax is in part the question whether and under what circumstances "a" or "the" syntax can be "read off" a semantic framework.

The issue is contested among philosophers of language.[41] There is the view that syntactic features are completely arbitrary and independent of semantic ones; and there is the view that syntactic features can always be recovered from semantic ones. O. Magidor has taken an intermediate position: "Some syntactic features, such as categorisations of words into 'noun' or 'verb', can be largely though not entirely predicted from semantic values",[42] as she has argued in [164].

In [163] Magidor has argued that sentences can be syntactically ill-formed but nevertheless meaningful. In foundational practice the reverse position, so to speak, is just what is known as (a version of) the syntactic point of view. This is the idea of a formalism in the radical sense of being blind or devoid of content – just as mathematics itself is asserted to be devoid of content, or in Gödel's terminology, the idea that mathematics may be viewed as a syntax of language.[43] Clearly the notion of semantic content meant here is different from

---

[40] The transient use of syntax also appears in set theory. See for example the notion of an "aa-mouse" in "Inner models from extended logics: Part 2", Kennedy, Magidor, Väänänen, manuscript.

[41] See Section 1.1.

[42] O. Magidor, personal communication.

[43] The view is laid out in detail in Gödel's "Is mathematics a syntax of language", 1953/9* in [87].

the idea of semantic content of a proposition of natural language. Nevertheless the question whether syntactic well-formedness must have or in some way give rise to semantic content, and the broader question of whether syntax is or can be wholly independent of semantics, has a clear meaning for foundational practice.

As for the larger question of when a semantically presented or natural language object or construction has, not necessarily an implicit logic, but beyond logic an implicit *formalisation*, this is simply the question of the adequacy of our foundational formal systems. The question is handled implicitly here and at length in the foundations of mathematics literature, mostly with respect to specific theories, e.g. in Baldwin's [10] via the notion of a local foundation for a specific topic area.

Here the question is conceptualised also on a level prior to full formalisation, so on the level of syntax and logic, as opposed to the level of axioms. Logicians have devised a number of ways of reading a syntax and logic off a semantic framework. We have seen how an AEC can give rise to a logic in the form of the Presentation Theorem, conjuring a language and a logic out of seemingly thin air. We remarked in Section 6.1 that the use of Skolem functions, in the original proof, or Skolem-like relations in the Baldwin–Boney proof, introduces arbitrariness and in that sense challenges the idea that the logic is in any way implicit to the class; while the reduction to an EC class rather than a PC class in the Shelah–Villaveces proof brings us closer to the notion of an implicit logic, one in which no extra predicates are needed.

There is also the more basic question how to conjure syntax from very basic initial data, for example from, simply, a model class, or indeed a game. We treated the latter in Section 2.2.3. We now turn to the former question. As it turns out, the logician always has a move to make, in the sense that one can always assign a syntax to a model class.

6.3.1 Extracting the Syntax from a Model Class

Given a model class with no (apparent) syntax, one can extract a syntax for the class by means of a generalised quantifier.[44]

To this end, suppose K is a model class. For simplicity assume that the vocabulary (i.e. similarity type) of K consists of just one binary predicate $P(x, y)$. Thus K is a class of structures of the form (M, P), where $P \subseteq M \times M$, and K is closed under isomorphism.

---

[44] See [265] for a nice introduction to the topic of generalised quantifiers.

Now add to first order logic the following generalised quantifier (in the sense of Lindström [152]):

$$M \models Q_K xy\phi(x, y, \vec{a}) \iff (M, \{(b, c) \in M^2 : M \models \phi(b, c, \vec{a})\}) \in K.$$

Observe that K is now definable in the extension $L(Q_K)$ of first order logic by the generalised quantifier Q_K:

$$K = \{(M, R) : (M, R) \models Q_K xy P(x, y)\}.$$

Moreover, $L(Q_K)$ is the smallest extension of first order logic to a logic in which K is definable. Why? Suppose $\phi(P)$ defines K in an extension L^* of first order logic. Then

$$Q_K xy\psi(x, y, \vec{z}) \leftrightarrow \phi(P(u, v)/\psi(u, v, \vec{z})).$$

Thus, if L^* is closed under substitution, $L(Q_K) \subseteq L^*$.

Given a whole family $K_i, i \in I$, of model classes – that is to say, a set of "sentences" of a logic in the semantic sense – we can add all the quantifiers $Q_{K_i}, i \in I$, to first order logic. In this way Lindström's logic (as in [153]), which has no a priori syntax, can be viewed as an extension $L(Q_i)_{i \in I}$ of first order logic with an explicit (albeit trivial) syntax.

In Section 2.2.1 we asked the question whether there is always implicit, ineliminable syntactic content in semantic frameworks (and conversely the question whether there is always an underlying semantic content in syntactic frameworks.) Here we recover a syntax from an arbitrary model class, by adding a quantifier to first order logic. Of course this new logic is given semantically; moreover we can only find the logic in question if we know the semantics already.

The method yields a syntax (of a kind) from a model class; however the method gives no clue as to what would be a natural way to axiomatise the new quantifier. In certain cases this can be done, for example, if K is the class of models (M, R), where $R \subseteq M$ is uncountable, then the extension of first order logic by the quantifier Q_R can be completely axiomatized by the so-called Keisler's Axioms [121] discussed in Section 5.4:

1. Axioms of first order logic
2. $\neg Q_R x(x = y \lor x = z)$
3. $\forall x(\phi \to \psi) \to (Q_R x\phi(x) \to Q_R x\psi(x))$
4. $Q_R x\phi(x) \to Q_R y\phi(y)$
5. $Q_R x \exists y\phi(x, y) \to (\exists y Q_R x\phi(x, y) \lor Q_R y \exists x\phi(x, y))$
6. Rules: Modus Ponens and Generalisation.

It can also happen that the extension of first order logic by the new quantifier Q_K has no Completeness Theorem with respect to an effectively given axiomatisation. For example, consider the case of the class K of models (M, R), where $R \subseteq M$ is infinite. Then if the sentence

$$\forall x \neg Q_K y(y < x)$$

is added to the usual (i.e. Peano's) first order axiomatisation of the natural numbers, the new theory T has up to isomorphism only one model, the standard one. But now it is clear that there cannot be any Completeness Theorem with respect to an effectively given axiomatisation for the quantifier Q_K, as otherwise the predicate "$T \vdash \phi$" would be recursively enumerable. But if this in turn were the case we could define arithmetic truth, as for any arithmetical sentence θ, θ is true in the standard model iff $T \vdash \phi$. This contradicts Tarski's Theorem, namely the fact that truth in the natural numbers is not arithmetically definable.

A drawback of the above method is that we obtain a different logic for each model class, in contrast to McGee's result. On the other hand as we noted, McGee's result depends on the cardinality of the models in the class in the sense that for each cardinality we obtain a separate definition.

A final point: if one thinks of a class in the usual sense of being definable in set theory, then the syntax of the class will be inherited from that of set theory.

6.4 A Remark on Set Theory

Baldwin's [10] follows the history of the entanglement of model theory with set theory, and the long and complex story subsequently, of how first order model theory detached itself from set theory. The "divorce", in Baldwin's terminology, of set theory and first order model theory is part of the paradigm shift he identifies, and is hinted at in the above remark of Sacks (see Section 2.2.1), that model-theoretic concepts are absolute, ergo without set-theoretic content – a remark we would certainly question. For example, Baldwin tells a detailed story in [10], how the analysis of infinitary and strong logics revealed the set-theoretic content of those logics.

To the extent that one restricts oneself to certain first order cases, then, one can perhaps speak, as Baldwin does, of a "divorce" between model theory and set theory.[45] In this book we have considered a broader class of logics, with

[45] [10], p. 170.

no turning away from set-theoretic content. Tarski seems to have approached model theory this way:

> The tendency to distinguish between those constructions and derivations which involve set-theoretical notions from those which do not involve them is undoubtedly a pronounced trend of modern algebraic research. The theory of arithmetical classes provides this trend with a theoretical framework and exhibits its reach and limitations ...
>
> The fact that we shall concern ourselves with those algebraic notions which involve no set-theoretical elements [i.e. notions arising from f.o. logic] by no means implies that we shall avoid set-theoretical apparatus in studying these notions. On the contrary, it will be seen that set-theoretical constructions and methods play an essential part in the development of the general theory of arithmetical classes.[46]

In the next section we will see one way in which set-theoretical constructions and methods play an essential part in the development of the general theory of arithmetical classes. In the framework given below, logics are reflected into natural language in a particular way; and then the semantic object is retrieved from natural language *along with some inherited new properties*. The treatment of logics here is reminiscent of that in the extended constructibility setting of the previous chapter, in which logics were varied in the construction of L in order to test for structural invariance. Below we give a second systematic framework, in which we treat logics parametrically in the search for their set-theoretic content.

6.5 Symbiosis

One of the aims of this book is to present calculi for measuring the entanglement of our natural language discourse with canonical formal languages. Here we offer an analysis of the entanglement of logics with natural set-theoretic language. We view this entanglement through the lens of *model classes*, so working firmly in the Tarskian vein.

Simple set-theoretical predicates (such as countability) are considered on their own. This is as opposed to considering more complex concepts, such as definability. The methodology here elaborates Tarski's remark that "set-theoretical constructions and methods play an essential part in the development of the general theory of arithmetical classes".[47]

The search for invariants of our natural language yields results of the kind: the invariance, or more precisely, the absoluteness of a logic guarantees the

[46] [249].
[47] [249].

invariance (or absoluteness) of the predicates with which it is entangled. Instead of entanglement, in this case we will use the word *symbiosis*, in keeping with the literature.

Symbiosis concerns the tight relationship between set-theoretic definability and the notion of a model class.[48] It was developed by Jouko Väänänen in order to, as he puts it, "expose the nature of the logic"; to "uncover the set-theoretical commitments of the logic, its content, its strength, even its reference".[49]

The relevance of symbiosis for us is that it provides a finely grained machinery for measuring a certain kind of entanglement, the entanglement of a *logic* with a concept of set theory such as "x is countable", "x is finite", "x is a cardinal number", "x is the power-set of y". With symbiosis we are able to detect whether a logic "sees" the invariant content of a given set-theoretic predicate – recognises, one might even say, its meaning. And on the other hand the absoluteness of the logic is pinned to the absoluteness of the predicate – whence the name "symbiosis".

Recent debates concerning the comparative virtues of second order logic vs. set theory, for example, depart from Tarski's explicit cosmopolitanism in decrying the entanglement of set theory with second order logic – insofar as it is admitted to exist at all.[50] Whereas from the symbiosis point of view, we will see (indeed prove) that second order logic is actually symbiotic with the power set operation.[51] Or to put it another way: *it is useless to try to separate second order logic from set theory.*

We recall some terminology: a logic is said to be absolute if the satisfaction predicate is Δ_1, and the property of being a formula of the logic is Σ_1 in the Levy hierarchy.[52] Intuitively, a logic $\mathcal{L}$ is absolute if the truth of a sentence in an $\mathcal{L}$-structure is not dependent on the background set theory. That is, the truth of a sentence depends only on the elements of the domain, not on what kind of subsets it has.[53] First order logic is absolute, as we noted, but there are other absolute logics such as $L_{\omega_1,\omega}$. Second order logic is famously non-absolute.

---

[48] A model class is simply a class of structures closed under isomorphism. Elementary classes are model classes, but not conversely, as there is no restriction to first order logic in the definition of model class.

[49] See [126].

[50] See for example [218]. On the popularity of second order logic among philosophers see C. Parsons [192], in which Parsons remarks: "In recent years there has been what might be called a love affair between philosophers of mathematics and second order logic."

[51] Baldwin [10] gives a thorough analysis of the entanglement of infinitary logics with set-theoretic assumptions.

[52] See [16], pp. 311–312.

[53] [16], ibid.

Symbiosis applies to a very broad class of logics but it was originally developed in order to answer the question: what blocks second order logic (henceforth SOL) from being absolute? We will see that the answer to the question lies in the fact that second order logic is symbiotic with the power set operation. One would expect the non-absoluteness of SOL to be tied to the power set operation, *somehow*. Symbiosis eliminates the "somehow".

We now define symbiosis. We require an auxiliary concept "R-absoluteness", defined as follows:

Definition 6.5.1 Suppose R is a predicate in the language $\{\in\}$. A predicate P is "absolute with respect to R", or "R-absolute" if it is absolute with respect to transitive extensions preserving the predicate R, i.e. if the predicate is preserved by extensions of the universe as long as R itself is preserved, and no new elements are added to old elements (technically: extensions of transitive models); and the same is true of restrictions of the universe. Technically this is the same as P being $\Delta_1(R)$ i.e. Δ_1 in the extended language $\{\in, R\}$.

Intuitively, a predicate P is R-absolute if whenever we add sets to the universe or take sets away, without changing R, also P remains unchanged. For example, if $R(x)$ is the predicate "x is countable", then the predicates "x is uncountable", "x is a countable ordinal", "x is a countable set of singletons", "$(A, <)$ is a linear order in which every initial segment is countable", "G is a graph in which every node has uncountably many neighbours", are all R-absolute. The predicate R is in the central role, somewhat analogous to that of an oracle: if R is preserved then P is preserved.

Recall that a model class is a class of structures of a fixed vocabulary closed under isomorphism. In a natural way, every model class K gives rise to a predicate, namely the predicate "$x \in K$". Conversely, every predicate P gives rise to a model class, denoted K_P. This is defined as follows:

Definition 6.5.2 Suppose P is an n-ary predicate. The model class K_P consists of models $\langle M, E, a_1, \ldots, a_n \rangle$ isomorphic to some $\langle M', \in, a'_1, \ldots, a'_n \rangle$ such that M' is a transitive set and $P(a'_1, \ldots, a'_n)$ holds.

We now explain the meaning of symbiosis in the special case of second order logic and the operation with which it is symbiotic, namely the power set operation. Consider a sentence ϕ of second order logic and let $P(x)$ be the predicate "x is a model and ϕ is true in x". This is a genuine predicate of set theory, albeit more complicated than the predicates we considered above. Let R be the binary predicate "x is the power set of y". One can show that P is

R-absolute, independent of the choice of the second order sentence ϕ. *This means that for establishing the truth of ϕ in a model all that is needed is the absoluteness of R and of predicates which are R-absolute.*

Conversely, let P be a predicate which is R-absolute, and consider the model class K_P associated to P. Then K_P is second order definable.[54]

Symbiosis was invented in order to explain the non-absoluteness of second order logic; to answer the question, what keeps second order logic from being absolute? An exact answer can now be given: the above R is the reason why second order logic is non-absolute. Once we adopt R-absoluteness, that is to say once we hold the power set operation fixed, second order logic becomes absolute. On the other hand, second order logic "sees" the predicate R and can talk about it and everything else that is R-absolute, via its definable model classes.

Väänänen describes the symbiosis between second order logic and the power set operation thus:

> When we talk about mathematics based on the power-set operation in some vague sense, it seems immaterial whether we use second order logic or predicates that are R-absolute. For second order logic we have the formal counterpart in the formal second order language. Likewise, for R-absoluteness we have $\Delta_1(R)$-formulas which capture the R-absolute properties, if we want to use a formal concept.[55]

Of course the predicate "x is the power set of y", is absolute in this sense only with respect to a special class of models of set theory. But the point here was not to study the predicate on its own; the aim was to determine whether a given logic could "see" the predicate through its definable model classes.

We now turn to the general case. Symbiosis is defined roughly as follows:

Definition 6.5.3 An n-ary predicate R and a logic $\mathcal{L}^*$ are *symbiotic* if exactly the absolute with respect to R properties of models can be expressed in $\mathcal{L}^*$.[56]

More exactly:

Definition 6.5.4 An n-ary predicate R and a logic $\mathcal{L}^*$ are *symbiotic* if the following conditions are satisfied:

---

[54] A detailed proof is reproduced in the appendix of the present chapter.

[55] Väänänen, personal communication.

[56] Barwise's concept of an *absolute logic* is related to symbiosis but is not the same. See [16]. An absolute logic as defined by Barwise requires the satisfaction predicate to be Δ_1, but without extra predicates. In the generalisation of the concept introduced by Väänänen one adds the predicate P as a kind of "oracle".

1. Every $\mathcal{L}^*$-definable model class is absolute w.r.t. R.
2. Every model class which is absolute w.r.t. R is $\mathcal{L}^*$-definable.[57]

Technically, what symbiosis tells us about a logic $\mathcal{L}^*$ is that its truth predicate is "recursive" in the predicate P, in the generalised sense of being $\Delta_1(P)$.

Examples of Symbiosis
The following pairs $(\mathcal{R}, \mathcal{L}^*)$ are symbiotic:[58]

1. $\mathcal{R}$: Cd, i.e. the predicate "x is a cardinal".
   $\mathcal{L}^*$: $\mathcal{L}_{\omega\omega}(I)$, where $Ixy\varphi(x)\psi(y) \leftrightarrow |\varphi| = |\psi|$ is the Härtig quantifier.
2. $\mathcal{R}$: Cd
   $\mathcal{L}^*$: $\mathcal{L}_{\omega\omega}(R)$, where $Rxy\varphi(x)\psi(y) \leftrightarrow |\varphi| \le |\psi|$ is the Rescher quantifier.
3. $\mathcal{R}$: Cd
   $\mathcal{L}^*$: $\mathcal{L}_{\omega\omega}(W^{Cd})$, where $W^{Cd}xy\varphi(x,y) \leftrightarrow \varphi(\cdot,\cdot)$ is a well-ordering of the order-type of a cardinal.
4. $\mathcal{R}$: Cd, WI
   $\mathcal{L}^*$: $\mathcal{L}_{\omega\omega}(I, W^{WI})$, where $W^{WI}x\varphi(x) \leftrightarrow |\varphi(\cdot)|$ is weakly inaccessible.
5. $\mathcal{R}$: Rg, i.e. the predicate "x is a regular cardinal".
   $\mathcal{L}^*$: $\mathcal{L}_{\omega\omega}(I, W^{Rg})$, where $W^{Rg}xy\varphi(x,y) \leftrightarrow \varphi(\cdot,\cdot)$ has the order-type of a regular cardinal.
6. $\mathcal{R}$: Cd, WC
   $\mathcal{L}^*$: $\mathcal{L}_{\omega\omega}(I, Q_{Br})$, where $Q_{Br}xy\varphi(x,y) \leftrightarrow \varphi(\cdot,\cdot)$ is a tree order of height some α and has no branch of length α.
7. $\mathcal{R}$: Cd, WC
   $\mathcal{L}^*$: $\mathcal{L}_{\omega\omega}(I, \bar{Q}_{Br})$, where $\bar{Q}_{Br}xyuv\varphi(x,y)\psi(u,v) \leftrightarrow \varphi(\cdot,\cdot)$ is a partial order with a chain of order-type $\psi(\cdot,\cdot)$.
8. $\mathcal{R}$: Pw, i.e. the predicate $\{(x,y) : y = \mathcal{P}(x)\}$
   $\mathcal{L}^*$: The second order logic $\mathcal{L}^2$.

There are many consequences of symbiosis, i.e. many logical concepts such as decision problems and so on have natural set-theoretic characterisations based on appropriate symbioses. For example the following is a kind of downward Löwenheim–Skolem theorem for logics $\mathcal{L}^*$. Consider the following *structural reflection principle* due to Joan Bagaria, a reflection principle for classes of models definable in a Σ_1 way from a predicate R of set theory:

---

[57] More precisely still, $\Delta(\mathcal{L}^*)$-definable. $\Delta(\mathcal{L}^*)$ is the logic the definable model classes of which are such K that both K and $-K$ are reducts of $\mathcal{L}^*$-definable model classes. See the Appendix for details.

[58] From [8].

$SR_R(\kappa)$: If K is a $\Sigma_1(R)$ class of models, then for every $M \in K$, there exist $N \in K$ of cardinality less than κ and an elementary embedding $e : N \preceq M$.

J. Bagaria and J. Väänänen [8] proved the following theorem about structural reflection:

Theorem 6.5.5 ([8]) *If R and $\mathcal{L}^*$ are symbiotic, the reflection principle for R and the Löwenheim–Skolem–Tarski theorem for $\mathcal{L}^*$ hold for the same cardinals.*

Symbiosis is a way of seeing that the content of a certain informal predicate (of set theory) can be tracked by a logic. This meant thinking of the concept of e.g. a well-ordering or of a cardinal or of the power set operation, as a natural language concept. The methodology is both set- and model-theoretic, and centralises the idea of a model class.

Symbiosis makes explicit the set-theoretical commitments (well-ordering, countability, cardinality, power set) of a logic. And on the other hand symbiosis tells us about the logical entanglements of an informal predicate in the following sense: the predicate determines the absoluteness of the logic, i.e. the logic becomes absolute once we know the predicate is. The logic and the predicate are entangled in a special sense.

The step taken here was to see the invariant content of a concept or predicate, in the informal sense its *meaning*, reflected in the concept of absoluteness, the absoluteness of the predicate with respect to transitive models of set theory.

6.5.1 The Road to Content: Invariance, Meaning and Context Relativism

We mean here informal content, and indicated two alternative roads in this chapter, in which the fundamental concept, the starting point in both cases, is the Tarskian notion of a model class. First, informal content can be captured by squeezing arguments. Originally the methodology of squeezing arguments involved proving the extensional equivalence of a chain of inclusions, trapping an informal concept in the centre of the chain, via a completeness theorem. In our "Tarskian" squeezing arguments extensional equivalence did not depend on a completeness theorem per se.

It appears that squeezing arguments can be broadly applied, if formulated in the right way. In [129] they were used to calibrate the escalating set-theoretical commitments of various logics. The (Kreiselian) squeezing argument for the logic $L(Q_1)$, for example, goes like this: Let $D_{L(Q_1)}$ denote the concept of formal provability relative to this logic. Let $V_{L(Q_1)}$ denote the truth of $L(Q_1)$-statements in all set-theoretical structures. Finally let $Val_{L(Q_1)}$ stand for the validity of $L(Q_1)$-statements relative to all possible structures. Then if the inclusions

$$D_{L(Q_1)} \subseteq Val_{L(Q_1)} \subseteq V_{L(Q_1)}$$

hold, the squeezing argument relative to the logic $L(Q_1)$ must also hold, by the completeness theorem for $L(Q_1)$.

We also cited *symbiosis* as a way of tracking the invariant content of an informal set-theoretical predicate. Proceeding in a mathematically direct way and with the "right" logic, the logic with which the predicate is symbiotic, we track a certain kind of invariance. On the other hand symbiosis provides a way of understanding the non-absoluteness of a logic.

Quine declared famously in [204] that "Second order logic is set theory in sheep's clothing." Väänänen has written this rejoinder [267]:

> The curious quality of second order logic is that the truth of a sentence such as Θ_{CH} in a big enough model of the empty vocabulary is equivalent to the truth of the Continuum Hypothesis in the set-theoretical universe. By the same technique almost any set-theoretical statement can be turned into a sentence of second order logic, the truth of which in a big enough model of the empty vocabulary is equivalent to the truth of the statement in the set-theoretical universe. Somehow second order logic manages to read the background set theory correctly. Does this mean that second order logic is set theory in "sheep's clothing"?

Symbiosis adds to what we know about the entanglement of set theory with second order logic, by providing a way of calibrating the (second order) logical content of the power set relation, and conversely.[59]

From the perspective of symbiosis the "second order logic vs set theory debate" was based on a binary opposition that was too strict. A silver lining here is that Quine's provocation may have invited logicians to take a closer look at second order logic – in the context of set theory and otherwise. We now know, for example, that most of the structures mathematicians work with are second order characterisable, and that it is very hard to find any such structure that is not second order characterisable.[60]

---

[59] In fact there are other calibration schemas. For example there is now a whole range of logics calibrated by (this time) large cardinals, in the sense that the assumption of the cardinal is equivalent to or implies a Löwenheim–Skolem–Tarski theorem for the logic. See e.g. [263].

[60] [266].

It is hardly a new observation that invariance, absoluteness, robustness and the like lay down a road to content. Some will take the view that there can be *no* context independent notion of semantic content in natural language.[61] The moderate contextualist claim that occasion-sensitivity abounds in natural mathematical language, is one we readily accept. At the same time, we are also able to uncover invariance in our foundational discourse, by means of calculi developed for the purpose. Symbiosis gave us one calculus for tracking the invariant content of a concept or predicate, or in the informal sense its *meaning*, via the absoluteness of the given predicate with respect to transitive models of set theory. Extended constructibility gave us another calculus, allowing us to test the sensitivity of a concept such as constructibility (among other concepts), to the various logics with which it is formalised. We leave the detailed analysis of how these calculi assign content to the concepts in question to future work. We merely wish to point out that foundational practice provides ample material for studying the question of content.

6.6 Coda: Symbiosis in Detail

6.6.1 Preliminaries

In this coda we will prove that second order logic is symbiotic with the power set operation. The proof is adapted from [261], in which the concept of symbiosis is presented in full generality, for absolute as well as non-absolute logics.

First we need the concept of relativised reduct. To this end, let $\mathcal{M}$ be a model with vocabulary L and $L_1 \cup \{P\} \subseteq L$, where P is unary. Then the *relativised reduct* $\mathcal{M}^P \upharpoonright L_1$ of $\mathcal{M}$ to the vocabulary L_1 and to the predicate P is the model $\mathcal{M}_1$ with $P^{\mathcal{M}}$ as the domain and the following structure:

- $R^{\mathcal{M}_1} = R^{\mathcal{M}} \cap (P^{\mathcal{M}})^n$ if $R \in L_1$ is of arity n.
- $f^{\mathcal{M}_1} = f^{\mathcal{M}} \upharpoonright (P^{\mathcal{M}})^n$ if $f \in L_1$ is of arity n.
- $c^{\mathcal{M}_1} = c^{\mathcal{M}}$ if $c \in L_1$.

The scalar field of a vector space V is an example of a relativised reduct, as is the additive group of the vectors of V.

We now extend the above definition of the relativised reduct of an individual model to the relativised reduct of a *class* of models as follows: If K is a model

---

[61] See, e.g., Travis, [256].

class with vocabulary L, and $L_1 \cup \{P\} \subseteq L$, then $K^P \upharpoonright L_1$ is the class of relativised reducts $\mathcal{M}^P \upharpoonright L_1$ of models $\mathcal{M} \in K$.

Thus we can obtain from, e.g., a class of vector spaces, their relativised reduct: a class of (scalar) fields, a class of (additive) groups, and a class of multiplicative groups.

We can now define our principal auxiliary technical notion, that of a Δ-extension:

Definition 6.6.1 ([17]) The Δ-extension $\Delta(\mathcal{L}^*)$ of a logic $\mathcal{L}^*$ is the logic the definable model classes of which are such K that both K and the complement of K are relativised reducts of $\mathcal{L}^*$-definable model classes.

$\Delta(\mathcal{L}^*)$ is equivalent to $\mathcal{L}^*$ for some logics, e.g. for first order logic and for the infinitary logic $L_{\omega_1\omega}$. However often there are properties of models that the logic $\mathcal{L}^*$ cannot express, but for trivial reasons, while they are expressible in $\Delta(\mathcal{L}^*)$. This anomaly is repaired by taking the Δ-extension. As an example, recall that the logic $L(Q_0)$ denotes first order logic with the generalised quantifier "there are infinitely many", and L_w^2 denotes weak second order logic with second order quantifiers for second order variables that range over finite subsets and relations. Then $\Delta(L(Q_0)) \equiv \Delta(L_w^2)$, i.e. the logics $\Delta(L(Q_0))$ and $\Delta(L_w^2)$ have the same definable model classes [17]. However the two logics are not equivalent, as $L(Q_0)$ is strictly stronger than L_w^2. See [165] for properties of the Δ-operation.

The Δ-operation is a closure operation in the sense that for all logics $\mathcal{L}^{*}$[62] [165]:

1. $\mathcal{L}^* \leq \Delta(\mathcal{L}^*)$. (Increasing)
2. $\mathcal{L}^* \leq \mathcal{L}^{**}$ implies $\Delta(\mathcal{L}^*) \leq \Delta(\mathcal{L}^{**})$. (Monotone)
3. $\Delta(\Delta(\mathcal{L}^*)) \equiv \Delta(\mathcal{L}^*)$. (Idempotent)

Moreover, the Δ-operation preserves the essential properties of a logic [165]:

1. If every sentence of $\mathcal{L}^*$ which has a model has also a countable model, then the same is true of every sentence of $\Delta(\mathcal{L}^*)$. (Löwenheim–Skolem Property)
2. If every set of sentences of $\mathcal{L}^*$, every finite subset of which has a model, itself has a model, then the same is true of $\Delta(\mathcal{L}^*)$. (Compactness Property)
3. If there is a recursive complete axiomatisation for (validity in the logic) $\mathcal{L}^*$, then the same is true of $\Delta(\mathcal{L}^*)$. (Completeness Property)

---

[62] If $\mathcal{L}^*$ and $\mathcal{L}^{**}$ are logics, we write $\mathcal{L}^* \leq \mathcal{L}^{**}$ if every model class definable in $\mathcal{L}^*$ is also definable in $\mathcal{L}^{**}$. We write $\mathcal{L}^* \equiv \mathcal{L}^{**}$ if $\mathcal{L}^* \leq \mathcal{L}^{**}$ and $\mathcal{L}^{**} \leq \mathcal{L}^*$.

This is the reason why $\Delta(\mathcal{L}^*)$ is considered a "minor" modification of $\mathcal{L}^*$.

6.6.2 Symbiosis Defined

Here is the technical definition of symbiosis:

Definition 6.6.2 An n-ary predicate R and a logic $\mathcal{L}^*$ are *symbiotic* if the following conditions are satisfied:

1. Every $\mathcal{L}^*$-definable model class is absolute with respect to R.
2. Every model class which is absolute with respect to R is $\Delta(\mathcal{L}^*)$-definable.

Note that the Δ-extension of a logic is given semantically, and there is no presence of a syntax. For the logic $L(Q_0)$ the (or a) syntax of $\Delta(L(Q_0))$ is known, while in the case of $\Delta(L(Q_1))$ no syntax has been found as yet, and similarly with second order logic L^2, no syntax is known for $\Delta(L^2)$.

6.6.3 Second Order Logic

We shall now establish symbiosis in the important special case of second order logic. Let Pw be the relation $\text{Pw}(x, y) \leftrightarrow y = \mathcal{P}(x)$. Let $\mathcal{SO}$ be second order logic.

Theorem 6.6.3 ([261], [8]) *The relation* Pw *and the logic* $\mathcal{SO}$ *are symbiotic.*

Proof We need to prove the following

1. Every $\mathcal{SO}$-definable model class is $\Delta_1(\text{Pw})$-definable.
2. Every $\Delta_1(\text{Pw})$-definable model class is $\Delta(\mathcal{SO})$-definable.

To statement 1, suppose $\mathcal{K} = Mod(\varphi)$, for some $\mathcal{SO}$-sentence φ. Then $\mathcal{A} \in \mathcal{K}$ if and only if $\langle V_\alpha, \in, \mathcal{A} \rangle \models ``\mathcal{A} \models \varphi"$, for some (any) α strictly greater than the smallest β such that $\mathcal{A} \in V_\beta$. Since the predicate $x = V_\alpha$ is $\Delta_1(\text{Pw})$, $\mathcal{K}$ is $\Delta_1(\text{Pw})$-definable.

To statement 2, suppose $\mathcal{K}$ is a $\Delta_1(\text{Pw})$-definable model class that is closed under isomorphisms. Suppose the vocabulary of $\mathcal{K}$ is L_0 which we assume for simplicity to have just one sort s_1 and one binary predicate P. Let $\Phi(x)$ be a $\Sigma_1(\text{Pw})$-formula of set theory such that $\mathcal{A} \in \mathcal{K}$ if and only if $\Phi(\mathcal{A})$. Let s_0 be a new sort, E a new binary predicate symbol of sort s_0, F a new function symbol from sort s_1 to sort s_0, and c a new constant symbol of sort s_0. Consider the class $\mathcal{K}_1$ of models

$$\mathcal{N} = (N, B, E^{\mathcal{N}}, c^{\mathcal{N}}, F^{\mathcal{N}}, P^{\mathcal{N}}),$$

where N is the universe of sort s_0 and B the universe of sort s_1, that satisfy the sentence φ given by the conjunction of the following sentences of second order logic:

(i) The conjunction θ of a finite part of the first order axiom system ZFC^- (ZFC axioms except the power-set axiom) written in the vocabulary $\{E\}$ (instead of $\in$) in sort s_0. This finite part of ZFC^- is chosen so that everything that is needed is included.

(ii) The first order sentence $\Phi(c)$ written in the vocabulary $\{E\}$ in sort s_0.

(iii) The second order sentence which says that E is well-founded.

(iv) The following second order sentence: $\forall x \forall W (\forall y (W(y) \rightarrow yEx) \rightarrow \exists z \forall y (W(y) \leftrightarrow yEz))$.

(v) A first order sentence saying that c is a pair (a, b), where $b \subseteq a \times a$, all written in the vocabulary E in sort s_0.

(vi) A first order sentence saying that F is an isomorphism between the s_1-part $(B, P^{\mathcal{N}})$ of the model and the structure $(\{x \in N : xE^{\mathcal{N}}a\}, \{(x, y) \in N^2 : (x, y)E^{\mathcal{N}}b\})$.

Claim: $\mathcal{A} \in \mathcal{K}$ if and only if $\mathcal{A} = \mathcal{N} \upharpoonright \{s_1, P\}$, for some $\mathcal{N} \in \mathcal{K}_1$.

Suppose first $\mathcal{A} \in \mathcal{K}$. Note that Pw is Σ_2-definable. Let n be such that θ is a Σ_n-sentence. By the Levy Reflection Principle ([113, Theorem 7.4]), let $V_\alpha \preceq_{n+2} V$, with $\mathcal{A} \in V_\alpha$. Then, $\mathcal{A} = \mathcal{N} \upharpoonright \{s_1, P\}$, where

$$\mathcal{N} := (V_\alpha, A, \in, \mathcal{A}, id, P^{\mathcal{A}}) \in \mathcal{K}_1.$$

Conversely, suppose $\mathcal{N} := (N, B, E^{\mathcal{N}}, c^{\mathcal{N}}, F^{\mathcal{N}}, P^{\mathcal{N}}) \in \mathcal{K}_1$ with $\mathcal{A} = \mathcal{N} \upharpoonright \{s_1, P\}$. Clearly, the structure $(N, E^{\mathcal{N}})$ is extensional and well-founded (by condition (iii)). Moreover, $(N, E^{\mathcal{N}}) \models \Phi(c^{\mathcal{N}})$. Since $\mathcal{K}$ is closed under isomorphisms, we may assume, without loss of generality, that N is a transitive set and $E^{\mathcal{N}} = \in \upharpoonright N$ (we are appealing here to the so-called Mostowski's Collapsing Lemma [113, page 65]). Now, Pw is absolute for N: for every a, b in N, we know that Pw(a, b) iff $N \models$ Pw(a, b). Since $(N, \in) \models \Phi(c^{\mathcal{N}})$ and N is transitive, and since Φ is Σ_1(Pw), we have that $\Phi(c^{\mathcal{N}})$ is true, i.e., it holds in V. Thanks to condition (vi), $c^{\mathcal{N}}$ is a binary structure isomorphic to $\mathcal{A}$. Since $\mathcal{K}$ is closed under isomorphism, $\mathcal{A} \in \mathcal{K}$. Thus $\mathcal{K}$ is a projection of the $\mathcal{SO}$-definable model class $\mathcal{K}_1$. The Claim is proved.

We can do the same for the complement of $\mathcal{K}$. Hence $\mathcal{K}$ is $\Delta(\mathcal{SO})$-definable. Statement 2 is proved.

$\square$

7

On the Side of Natural Language

We opened our 2013 [124] with the remark that mathematics is practised in a formalism free manner, that this has always been the case, and will remain the case – formalism freeness meaning, in part: carried out in natural language with no mention of syntax and semantics, never mind that a distinction might be drawn between the two. The discipline known as foundations of mathematics is practised differently: the foundationalist seeking a grounding for mathematical discourse finds it not in natural language but in formalisations of various kinds – in fact the foundationalist, in general, shies away from natural language.

Sagi took note of this way of thinking in a recent paper on the question (treated, e.g., by M. Glanzburg [78]) whether natural language has a logic at all, in particular whether natural language has a notion of consequence:

Logicians often approach natural language with some apprehension: natural language is complex and messy, studied fragment by fragment by a variety of methods that hardly seem to provide any sense of unity. This is by contrast to formal languages, that are neat, manageable and simple (to the extent that the logician devises them to be). Is there even a logic in natural language?[1]

It is not surprising that Tarski made the point as well in 1936, also regarding the divergence of formal and natural language consequence in his manuscript "On the concept of logical consequence": "the formalized concept of consequence, as it is generally used by mathematical logicians, by no means coincides with the common concept"[2].

[1] Sagi, "Considerations on logical consequence and natural language," manuscript.
[2] [251], p. 411. Tarski introduces an abstract consequence relation in the lecture. As described in Cintula et al.'s "An abstract approach to consequence relations," this was the, to the authors, "most successful" attempt to combine the syntactic and semantic features (of propositional logic) in a single neutral axiomatic framework. [39], p. 331.

In this book we sketched out a kind of hybrid foundational analysis, in which mathematical practice is considered on its own, that is to say *in situ* – in natural language – but checked by foundational practice in ways dictated by specific calculi. The first calculus considered here, of extended constructibility, involved viewing the concept of constructibility in set theory as a natural language concept, and then "testing" the concept against various logics. A second calculus involved the symbiosis between logics and set-theoretic operations considered in their natural language formulation. With these calculi in place we then studied the entanglement, or on the other hand the degree of formalism freeness inherent to the formal environments built around these concepts – entanglements with respect to formal systems lying largely between first and second order logic, though not exclusively.

We referred, en passant, to the semantic point of view. We suggested, following Tarski, that metamathematics can be and often is pursued from within the semantic point of view, in the sense that many essential metamathematical theorems are found to admit admit a semantic formulation, generally in terms of model classes, but not always – geometric and topological concepts may appear in the reformulation. As for the way mathematics itself is pursued we quoted a theorem to the effect that the semantic method enjoys a speed-up over the method of formal proof.[3] The theorem seems to say that the mathematician's inclination toward working semantically is sound, because the semantic method brings *efficiency*, in the sense of shortening the lengths of proofs.

One would be tempted, then, to speak of a mathematical methodology as being purely semantic; to speak of a proof, for example, as being a pure semantic proof. We have stood by the idea in this book, by and large. But we have also noted that semantic frameworks or concepts may be entangled, in the sense of the two calculi offered here: extended constructibility and symbiosis; but also in the sense of having an implicit syntax or logic. For example as we noted any model class gives rise not only to a syntax but to a logic, e.g. via McGee's theorem, or via generalised quantifiers. The theorem in question in both cases is easily proved – perhaps too easily proved.[4] Both proofs require the axiom of choice and thus the syntax and logic which one has read off the model class has a degree of arbitrariness. The improvement of McGee's theorem offered here for an arbitrary model class brings us closer to the idea of implicit logic, in that the logic defining the model class is well-behaved in various ways, in comparison to $L_{\infty\infty}$. (See Section 5.6.)

---

[3] See Section 1.2.

[4] Recall that to define a model class via McGee's theorem one simply lists everything that "happens" in the class, and this turns out to be a statement in the logic $L_{\infty\infty}$, for each cardinality separately.

We then turned to the question, not whether any model class has an implicit syntax (or logic), but whether a *well-behaved* model class has an implicit syntax (or logic). Shelah gave an answer for the AEC's in the form of the Presentation Theorem, showing that any AEC can be defined as a PC class, expanding the vocabulary of the class. Baldwin and Boney gave a different version of the theorem using "Skolem-like" relations, also expressing the AEC as a PC-class, so in an expanded vocabulary. Whereas Shelah and Villaveces's recent proof defines the AEC as an EC-class, that is, no extra predicates are needed, moving closer to the idea of an implicit syntax or logic.

It is possible that different notions of well-behavedness give rise to different logics. The first order structures are as well-behaved as can be, as current first order model theory attests – but then we have given up the idea of considering model classes in general.

We highlighted the dynamic aspect of the syntax/semantics distinction, and the fact that in important areas of model theory the distinction is perpetually redrawn. We also suggested that the *fragility* of the distinction was built into the practice from its inception; that however one wishes to classify the twin poles of syntax and semantics, they cannot be kept apart for long. This raises the question whether the syntax/semantics distinction is not occasionally unstable.

Seen in this light it is surprising that the syntax/semantics distinction is taken for granted, generally, in foundations of mathematics, in contrast to how the distinction is treated in other literatures. F. Moltmann has questioned the standard picture of how reference functions in the natural language use of number terms, as we noted; whether the word→object relation, or syntactic string→sortal relation, and other relations of this like – relations which are intimately bound up with the syntax/semantics distinction – function in the way we think they do, at least when it comes to ,"number talk".[5]

Also from the direction of philosophy of language, O. Magidor has problematised the syntax/semantics distinction, this time from the standpoint of the category mistake. Magidor has argued against the idea that category mistakes are syntactically well-formed but meaningless. Of course, for a different notion of meaning there is a sense in which this is precisely what the Hilbert programme tried to accomplish, but with mathematical discourse überhaupt, or alternatively with what is attempted by the syntactic point view: a reconstruction of mathematics as a syntax of language, but fully content free.

---

[5] On Moltmann's view, "numbers are not primarily abstract objects, but rather 'aspects' of pluralities of ordinary objects, namely number tropes …" [177], p. 2.

One cannot help thinking of the unfortunate Sublieutenant Zasetsky, who appears in Yuri Manin's logic textbook [170] (in the chapter called "Truth and deducibility"), and who suffers from asemia as a result of a head injury.[6] Manin, who uses the sublieutenant's predicament in order to challenge the epistemic value of formal derivations, is struck by the sublieutenant's oddly articulate descriptions of his disorder, the "complexity of Zasetsky's [contemporaneous JK] metalinguistic text describing his linguistic difficulties":

> Sometimes I'll try to make sense out of those simple questions about the elephant and the fly, decide which is right or wrong. I know that when you rearrange the words, the meaning changes. At first I didn't think it did, it didn't seem to make any difference whether or not you rearranged the words. But after I thought about it a while I noticed that the sense of the four words (*elephant, fly, smaller, larger*) did change when the words were in a different order. But my brain, my memory, can't figure out right away what the word *smaller* (or *larger*) refers to. So I always have to think about them for a while ... So sometimes ridiculous expressions like "a fly is bigger than an elephant" seem right to me, and I have to think about it a while longer.

For Manin, the fact that "the subtlety of the analysis seems incompatible with the crude errors being analyzed", indicates that there is a lesson to be drawn:

> All of this shows that there is no basis whatsoever for Rosser's opinion that "once the proof is discovered, and stated in symbolic logic, it can be checked by a moron." The human mind is not at all well suited for analyzing formal texts.[7]

Sublieutenant Zasetsky's contemporaneous writings are conceptually, syntactically and even, as Manin observes, *logically* complex. "I know that when you rearrange the words, the meaning changes" – but that is all Zasetsky knows, prior to recovering from his peculiar ailment.

In the eyes of the foundationalist, the mathematician and the sublieutenant are in the same boat, *philosophically*. They are both logically competent; but if asemia is marked by the inability to connect symbols with their meaning, then for the foundationalist the mathematician is asemic too – not in the sense of a basic linguistic competence of knowing the meanings of words but in a more profound sense of having no semantic theory, no theory of meaning—or at best having only a very attenuated one.[8]

---

[6] Asemia is a medical condition marked by the inability to understand the meanings of signs or symbols, i.e. the inability to connect symbols with their meaning.

[7] Manin, ibid, p. 36.

[8] As Bertrand Russell put it in 1917:

Is this the best we can do, as foundationalists? The labour of foundational analysis, insofar as it is directed toward the concept of inference, or consequence, or truth, or the like, often begins by dividing logical space into its syntactic and semantic modes. But mathematical language – or in our terminology, formalism free language – is marked, necessarily, by the very absence of this division. "Mathematical" discourse, in Tarski's sense, draws no such distinction, even as it may be refracted through that lens by others.

The neat division between formal theory and informal metatheory stands unchallenged by us. We do speak about formal theories in an informal metalanguage – an informal language which, with a slight shift of attention, can be formalised. The ability to shift one's attention in this way is a peculiar competence, an odd kind of bilingualism we referred to as *on-again-off-again-ism*, meaning the transient and opportunistic use of foundational formal systems in model theory and in mathematics proper. We were inspired to the terminology by Hodges's [106], in which he notices that in the Polish School the idea of "practising" deductive theories is even built into the language, through the phrase *uprawiać sformalizowane nauki dedukcyjne*, meaning, to "practise formalised deductive sciences".

Ordinary language philosophers see linguistic competence obtaining across the board, not the competence of mathematicians but the linguistic competence of *all* speakers of a language. To quote S. Laugier:

> The philosophy of language is interested in meaningfulness, mean-ing. Whether Carnap's or Austin's, this philosophy does not seek to establish a universal science of language that would lead us to discover, about or in language, something external to it: something that we don't know. For Carnap or Quine, the task is not to discover but to construct and invent; for Austin or Wittgenstein, everything is already there, and it remains for us to discover that we know it. Here lies a fundamental difficulty of the philosophy of language: the fact that our language (including philosophical) is always "unhappy" (the expression is Bouveresse's): "permanently haunted by a guilty conscience and a feeling of failure, never certain of its status and possibilities."[9]

The view taken in our Chapter 5 is that Tarski's conceptualisation of "the mathematical" treats the speaker of a mathematical language as knowing all

---

Pure mathematics consists entirely of assertions to the effect that, if such and such a proposition is true of anything, then such and such another proposition is true of that thing. It is essential not to discuss whether the first proposition is really true, and not to mention what the anything is, of which it is supposed to be true. ... Thus mathematics may be defined as the subject in which we never know what we are talking about, nor whether what we are saying is true. People who have been puzzled by the beginnings of mathematics will, I hope, find comfort in this definition, and will probably agree that it is accurate. [211], p. 50.

[9] Bouveresse, La parole malheureuse, 11. Translation S.L. [148], p. 113.

there is to know about that language – everything is already there, and it [only JK] remains to discover that s/he knows it as well. "Tarski takes his readers to understand the notion of 'mathematical property'", as Baldwin writes[10] — an assertion the foundationalist vigorously disputes. For foundationalist the natural language of mathematics is *profoundly unhappy*.

Our story began with the emergence of the concept of computability and ended with the idea that second order logic is *symbiotic* with the power set operation. In telling this particular story, in this particular way, we hoped to begin the process of unearthing the profile of natural language which is buried in foundational terrain. To highlight Gödel's lapse into natural language, and the same for Tarski's "run" toward toward natural language, as we called it – steps taken deliberately and for profoundly philosophical reasons. To put it another way, we wanted to investigate the standpoint in which one has an overview of foundational formal systems, without being entangled in any one of them.

Where are we? A link between the model-theoretic developments high-lighted here, and certain other lines of thinking, might be worth pointing out. The tendency to suppress various aspects of a formalism – the development in logic which inspires us here – has been accompanied on the philosophical side of things, particularly recently, by a sense of fatigue with a priorist or "first" philosophical analysis, to use P. Maddy's terminology, and the fore-fronting of philosophical naturalism,[11] or implicitly naturalistic approaches.[12] Naturalism, in particular the naturalist's recommendation to "track mathematical practice", advocates a withdrawal of the foundationalist critique as its very starting point.

The naturalist's suggestion to become absorbed in mathematical practice bifurcates along various lines – one can become absorbed in different ways, after all. Maddy's *second philosopher*[13] is bound by a particular view of mathematics' empirical applications, for example, which the methodological naturalist is, generally, not; while Wilson's view is tied differently to empirical applications. Cutting across all of these lines of thinking is a shared belief in the obsolescence of the foundational project. Franks puts it this way in his [74]:

Recent philosophical writing about mathematics has largely abandoned the *a priorist* tradition and its accompanying interest in grounding mathematical activity. The foundational schools of the early twentieth century are now treated more like

---

10 [10], p. 294.
11 See [159–161], [74].
12 See e.g. [7], [51] and [9].
13 See Maddy, ibid.

historical attractions than like viable ways to enrich our understanding of mathematics. This shift in attitudes has resulted not so much from a piecemeal refutation of the various foundational programs, but from the gradual erosion of interest in laying foundations, from our culture's disenchantment with the idea that a philosophical grounding may put mathematical activity in plainer view, make more evident its rationality, or explain its ability to generate a special sort of knowledge about the world.[14]

It seems glaringly obvious by now that any foundational scheme will deliver on some counts, but not on others – whence our logical pluralism, or to paraphrase Wilson, the recognition that descriptive strategies must draw on a prismatic conceptualisation of the target area. The point has been made over and over by logicians and philosophers, but we took it upon ourselves to make the point once more.

What then of theoretical unification, the problem of the border between theories? The suggestion here is that one can keep an eye on these borders, keep an eye on how "our descriptive vocabularies adjust their referential foci from one investigative context to the next", as Wilson puts it [285], by tracking our logical entanglements, or on the other hand by tracking the degrees of formalism freeness inherent to a particular context. As we attempted to show in this book, foundational terrain admits vast areas of invariance, if we look at things in the right way—invariance which can have the effect of assimilating these border areas into itself, if not of dissolving them altogether.

In contrast to the various kinds of logical pluralism discussed in the contemporary literature,[15] the pluralism espoused here stands on neutral ground with respect to most of the current debates in the field. And although the neutral point of view is not to be confused with the view from outside, it is true that philosophers looking for a full-blown theory of reference in natural mathematical language, or for a position on the ontological status of mathematical objects, or for any analysis of natural language going beyond a simple reliance on its meaning aspect, did not find it here – as much as our discussion seemed to call for it. Instead we buried our aspirations toward logical autonomy in mathematical ground – somewhat in the way Tarski did, perhaps – in the hope of realising those aspirations in the form of theorems. That said, we suggested in our Section 6.5.1 that invariance discovered by means of the parametric use of logics, may be useful for developing a theory of meaning for mathematical discourse, given the fact that some content is clearly context independent, or in our terminology *formalism free*.

---

[14] [74] p. 169.
[15] Beall and Restall's [18] is a recent classic.

We emphasised the fragility of the syntax/semantics distinction, not because we reject the distinction, it is rather that we think there is more to do if we want to set the syntax/semantics distinction on solid ground. Solid ground (for us) is tied up with invariance, as it is for mathematicians of every stripe, logical or otherwise, in one way or the other. This meant treating logics parametrically in specific ways, a methodology that enabled us to uncover invariance in particular settings. Thus in Section 4.4.2 instead of thinking about the adequacy of a formalism taken on its own, we wanted to test the persistence of e.g., the ZFC axiomatisation under the permutation of various logics (but one can carry this out for other canonical theories, conceivably).

We also wanted to think about definability in set theory in a way that took advantage of the lessons implicit in the development of computability in the 1930s. More precisely we wanted to think about definability in set theory in a way that took Gödel's and Post's view of those developments to heart: notions of definability and provability should be developed which are absolute in the way computability is. We saw that confluence and grounding were key elements in Gödel's conceptual analysis of computability. To that end we offered an implementation based on taking an informal notion of definability, in this case constructibility, and varying the underlying logic, in the search for confluence. We found that the constructible hierarchy L is not particularly sensitive to the underlying first order logic in the sense that a large class of logics can be substituted for first order logic in the construction of L, and the structure doesn't change. The same is true of the structure HOD, the hereditarily ordinal definable sets. This established confluence, but left grounding for another day.

We wanted, in short, to make progress also on the mathematical side of things – an aspiration that many of the founders of the modern era in logic have thought of as the purpose of foundational work.

Kreisel made an entrance here at various key points. We drew on his work on informal rigour, mainly; to a somewhat lesser extent we drew on his view of history – his alluding to the force of historical evidence in his writings, of "intuitive notions standing the test of time" [141], or of "what can be learned from the historical record" [142], or even of $T E N O S$, short for: "tested experience, not only speculation" (in e.g. [140]).

What Kreisel is getting at, of course, is *historicism*, and the idea that mathematics is very *old*. Historicism is the idea that the nature of a subject is entirely comprehended in its development[16]. Let us substitute the word "partly", or

---

[16] See [149].

even "substantially" for the word "entirely". In fact in the analytic project devoted to mathematics, whether it be aimed toward developing an ontology and epistemology for mathematics, or at providing some other kind of grounding; whether it is aimed at explaining why such a grounding is not needed, or even when it pronounces an obituary on Ontology, as H. Putnam did in his [202],[17] historicism, even in the weak sense, is nowhere to be found. Analytic philosophers of mathematics *are* making important contributions to philosophy, as historians of their own subject. At the same time it is simply the case that for the analytic project in philosophy of mathematics to succeed on its own terms, mathematics must be stripped of its history if it is to fall under the knife of philosophical analysis.

S. Gandon pushes against this conception of the analytic project, interpreting Wilson's *Wandering Significance* [284] as laying out a project "to reform philosophy of language by applying frameworks coming from the mathematical concept's evolution to meaning in general".[18] Kreisel also seems to resist this picture. What Kreisel wants to direct our attention to is the idea that what "the mathematical" has going for it, along with its applicability, its clarity and its other virtues, is its stability over the long-term.

"Meaning no longer travels as freely and easily through time and space as it used to do", F. R. Ankersmit mused, interestingly, in his monograph *Sublime Historical Experience* [5]. As for historicism:

> Historicism is one of the most caustic intellectual acids that has ever been prepared by Western civilization ... Religion, metaphysics, tradition, truth, all of them proved to be soluble in the acid of historicism. Historicism is even a stronger acid than logic, for logic can also be historicized with dramatic consequences for the pretensions of logic to offer timeless truths; and if one were to attempt the reverse and to rationalize history ... history will prove, once again, to be the stronger partner of the two.

Have we said anything about natural language in this book? About its phonetic system, its morphology, or its syntax? We have spoken about natural language defined negatively, that is, in terms of what it is not. That is to say, we have spoken about formalism freeness. To have a concept of formalism

---

[17] From Putnam's *Ethics Without Ontology*:
The invitation to give the Hermes Lectures at the University of Perugia ... provided me with an opportunity to formulate and present in public something I realized I had long wanted to say, namely that the renewed (and continuing) respect of Ontology (the capital letter here is intentional!) following the publication of W.V. Quine's "On What There Is" at the midpoint of the last century has had disastrous consequences for just about every part of analytic philosophy. ([202], p. 2.)

[18] Gandon, personal communication.

freeness is to have a concept of natural language, in particular to know natural language when we see it, even as we know at the very same time, that the division between formal and natural language is often. a matter of intention We know, as trained logicians, when we are speaking in the informal mode and when we are "speaking" from within a formalism. Formal systems do have their own very distinct modes of speech after all; and the tendency in foundational work has been to adopt that mode of speech, as a kind of protective barrier against, e.g. inconsistency, as well as in the service of other philosophical and foundational goals.

Is it alright?[19] If consistency is at all a concern anymore, then surely mathematicians know enough, by now, to stay out of logical trouble. To paraphrase Laugier, each speaker of a natural mathematical language already seems to know all there is to know about that language. Everything is already there, and it only remains for the foundationalist to discover that she knows it as well.[20]

So why not take a another look at the natural language of the mathematician – why not be on its side for a change?

[19] K. Manders asks this in Manders, K. Logic and Conceptual Relationships in Science (1985). *Logic colloquium '85 (Orsay, 1985)*, 193–211, Stud. Logic Found. Math., 122, *North-Holland, Amsterdam*, 1987.
[20] [148], p. 113.

Bibliography

[1] Wilhelm Ackermann. Zum Hilbertschen Aufbau der reellen Zahlen. *Math. Ann.*, 99(1):118–133, 1928.

[2] Hans Adler. Thorn-forking as local forking. *J. Math. Log.*, 9(1):21–38, 2009.

[3] Jyrki Akkanen. Absolute logics. *Ann. Acad. Sci. Fenn. Ser. A I Math. Dissertationes*, (100):83, 1995.

[4] Hajnal Andréka, Judit X. Madarász, István Németi, et al. A logic road from special relativity to general relativity. *Synthese*, 186(2):633–649, 2012.

[5] Frank R. Ankersmit. *Sublime Historical Experience*. Stanford University Press, Stanford, 2005.

[6] Emily Apter. *Against World Literature: On the Politics of Untranslatability*. Verso, 2013.

[7] Andrew Arana and Paolo Mancosu. On the relationship between plane and solid geometry. *Rev. Symb. Log.*, 5(2):294–353, 2012.

[8] Joan Bagaria and Jouko Väänänen. On the symbiosis between model-theoretic and set-theoretic properties of large cardinals. *J. Symb. Log.*, 81(2):584–604, 2016.

[9] John T. Baldwin. Formalization, primitive concepts, and purity. *Rev. Symbolic Logic*, 6(2):87–128, 2013.

[10] John T. Baldwin. *Model Theory and the Philosophy of Mathematical Practice*. Cambridge University Press, 2018. Formalization without foundationalism.

[11] John T. Baldwin and Will Boney. Hanf numbers and presentation theorems in AECs. In *Beyond First Order Model Theory*, pages 327–352. CRC Press, Boca Raton, FL, 2017.

[12] John T. Baldwin, T. Hyttinen and M. Kesälä. Beyond first order logic: From number of structures to structure of numbers part I. *Bulletin of Iranian Math. Soc.*, 2011. Available online at www.iranjournals.ir/ims/bulletin/ in press.

[13] John T. Baldwin, T. Hyttinen and M. Kesälä. Beyond first order logic: From number of structures to structure of numbers part II. *Bulletin of Iranian Math. Soc.*, 2011. Available online at www.iranjournals.ir/ims/bulletin/ in press.

[14] Jon Barwise. *Admissible Sets and Structures*. Springer-Verlag, Berlin-New York, 1975. An approach to definability theory, Perspectives in Mathematical Logic.

[15] Jon Barwise, Matt Kaufmann and Michael Makkai. Stationary logic. *Ann. Math. Logic*, 13(2):171–224, 1978.

[16] Jon Barwise. Absolute logics and $L_{\infty\omega}$. *Ann. Math. Logic*, 4:309–340, 1972.

[17] Jon Barwise. Axioms for abstract model theory. *Ann. Math. Logic*, 7:221–265, 1974.

[18] J. C. Beall and Greg Restall. *Logical Pluralism*. The Clarendon Press, Oxford University Press, 2006.

[19] Paul Bernays and Moses Schönfinkel. Zum entscheidungsproblem der mathematischen logik. *Mathematische Annalen*, 99(1):342–372, 1928.

[20] Arianna Betti. Polish axiomatics and its truth: on Tarski's Leśniewskian background and the Ajdukiewicz connection. In Douglas Patterson, editor, *New Essays on Tarski and Philosophy*, pages 44–71. Oxford University Press, 2008.

[21] Garrett Birkhoff. On the structure of abstract algebras. *Proc. Cambridge Phil. Soc.*, 31(7):434–454, 1935.

[22] Errett Bishop. *Foundations of Constructive Analysis*. McGraw-Hill Book Co., New York, 1967.

[23] Will Boney. Tameness from large cardinal axioms. *J. Symb. Log.*, 79(4):1092–1119, 2014.

[24] Denis Bonnay. Logicality and invariance. *Bull. Symbolic Logic*, 14(1):29–68, 2008.

[25] Denis Bonnay. Carnap's criterion of logicality. In *Carnap's Logical Syntax of Language*, pages 147–164. Macmillan, 2009.

[26] Denis Bonnay and Dag Westerstå hl. Compositionality solves Carnap's problem. *Erkenntnis*, 81(4):721–739, 2016.

[27] Nicholas Bourbaki. The architecture of mathematics. *Amer. Math. Monthly*, 57:221–232, 1950.

[28] L. E. J. Brouwer. *Brouwer's Cambridge Lectures on Intuitionism*. Cambridge University Press, 1981. Edited and with a preface by D. van Dalen.

[29] John P. Burgess. Descriptive set theory and infinitary languages. *Zb. Rad. Mat. Inst. Beograd (N.S.)*, 2(10):9–30, 1977. Set theory, foundations of mathematics (Proc. Sympos., Belgrade, 1977).

[30] John P. Burgess. Proofs about proofs: a defense of classical logic. In *Proof, Logic and Formalization*, pages 8–23. Routledge, London, 1992.

[31] John P. Burgess. *Mathematics, Models, and Modality*. Cambridge University Press, 2008.

[32] John P. Burgess. *Rigor and Structure*. Oxford University Press, 2015.

[33] Rudolf Carnap. *Der logische Aufbau der Welt*, volume 514 of *Philosophische Bibliothek [Philosophical Library]*. Felix Meiner Verlag, Hamburg, 1998. Reprint of the 1928 original and of the author's preface to the 1961 edition.

[34] Rudolph Carnap. *Formalization of Logic*. Harvard University Press, Cambridge, MA, 1943.

[35] Alonzo Church. A set of postulates for the foundation of logic. I, II. *Ann. Math. (2)*, 33:346–366, 1932.

[36] Alonzo Church. An unsolvable problem of elementary number theory. *Bull. Am. Math. Soc.*, 41:332–333, 1935.

[37] Alonzo Church. A note on the Entscheidungsproblem. *J. Symb. Log.*, 1:40–41 (Correction 1:101–102), 1936.

[38] Alonzo Church. The Richard Paradox. *Amer. Math. Monthly*, 41(6):356–361, 1934.

[39] Petr Cintula, José Gil-Férez, Tommaso Moraschini and Francesco Paoli. An abstract approach to consequence relations. *Rev. Symb. Log.*, 12(2):331–371, 2019.

[40] Dermot Moran and Joseph Cohen *The Husserl Dictionary*. Continuum, 2012.

[41] Paul Cohen. The independence of the continuum hypothesis. *Proc. Nat. Acad. Sci. U.S.A.*, 50:1143–1148, 1963.

[42] Jack Copeland and Diane Proudfoot. Deviant encodings and turing's analysis of computability. *Stud. Hist. Phil. Sci. Part A*, 41(3):247–252, 2010.

[43] Leo Corry. Nicolas Bourbaki and the concept of mathematical structure. *Synthese*, 92(3):315–348, 1992.

[44] Max J. Cresswell. *Logics and Languages*. Methuen & Co., Ltd., London, 1973.

[45] Martin Davis. *Computability and Unsolvability*. McGraw-Hill Series in Information Processing and Computers. McGraw-Hill Book Co., Inc., New York, 1958.

[46] Martin Davis. Why Gödel didn't have Church's Thesis. *Inform. and Control*, 54(1–2):3–24, 1982.

[47] Martin Davis, editor. *The Undecidable*. Dover Publications Inc., Mineola, NY, 2004. Basic papers on undecidable propositions, unsolvable problems and computable functions, Corrected reprint of the 1965 original [Raven Press, Hewlett, NY; MR0189996].

[48] Liesbeth De Mol. Closing the circle: an analysis of Emil Post's early work. *Bull. Symbolic Logic*, 12(2):267–289, 2006.

[49] Silvia De Toffoli and Valeria Giardino. Envisioning transformations – the practice of topology. In *Mathematical cultures*, Trends in the History of Science, pages 25–50. Birkhäuser/Springer, Cham, 2016.

[50] Walter Dean. Squeezing feasibility. In *Pursuit of the Universal (Lecture Notes in Computer Science)*, pages 78–88. Springer, 2016.

[51] Michael Detlefsen and Andrew Arana. Purity of methods. *Philosophers' Imprint*, 2011. Available online at www.philosophersimprint.org/.

[52] Michael Detlefsen. Poincaré vs. Russell on the rôle of logic in mathematics. *Philos. Math. (3)*, 1(1):24–49, 1993.

[53] Michael Detlefsen. What does Gödel's second theorem say? *Philos. Math. (3)*, 9(1):37–71, 2001. The George Boolos Memorial Symposium, II (Notre Dame, IN, 1998).

[54] Michael Detlefsen. Abstraction, axiomatization and rigor: Pasch and Hilbert. In *Hilary Putnam on Logic and Mathematics*, volume 9 of Outstanding Contributions to Logic, pages 161–178. Springer, Cham, 2018.

[55] Andrzej Ehrenfeucht. Application of games to some problems of mathematical logic. *Bull. Acad. Polon. Sci. Cl. III.*, 5:35–37, IV, 1957.

[56] Ronald Fagin. Probabilities on finite models. *J. Symb. Log.*, 41(1):50–58, 1976.

[57] Anita Burdman Feferman and Solomon Feferman. *Alfred Tarski: Life and Logic*. Cambridge University Press, 2004.

[58] Solomon Feferman. Arithmetization of metamathematics in a general setting. *Fund. Math.*, 49:35–92, 1960/1961.

[59] Solomon Feferman and Georg Kreisel. Persistent and invariant formulas relative to theories of higher order. *Bull. Amer. Math. Soc.*, 72:480–485, 1966.

[60] Solomon Feferman. Logic, logics, and logicism. *Notre Dame J. Form. Log.*, 40:31–54, 1999. Special issue in honor and memory of George S. Boolos.

[61] Solomon Feferman. Tarski's conceptual analysis of semantical notions. In Douglas Patterson, editor, *New Essays on Tarski and Philosophy*, pages 72–93. Oxford University Press, 2008.

[62] Solomon Feferman. Set-theoretical invariance criteria for logicality. *Notre Dame J. Form. Log.*, 51(1):3–20, 2010.

[63] Qi Feng, Menachem Magidor and Hugh Woodin. Universally Baire sets of reals. In *Set Theory of the Continuum (Berkeley, CA, 1989)*, volume 26 of Mathematical Sciences Research Institute Publications, pages 203–242. Springer, New York, 1992.

[64] Briony Fer. *On Abstract Art*. Yale University Press, 1997.

[65] Hartry Field. Tarski's theory of truth. *J. Philos.*, 69(13):347–375, 1972.

[66] Hartry Field. *Realism, Mathematics and Modality*. Basil Blackwell, Inc., New York, 1989.

[67] Hartry Field. *Saving Truth from Paradox*. Oxford University Press, 2008.

[68] Juliet Floyd. The varieties of rigorous experience. In *The Oxford Handbook of The History of Analytic Philosophy*, Oxford Handbooks in Philosophy. Oxford University Press, 2013.

[69] Juliet Floyd and Akihiro Kanamori. Gödel vis–vis Russell: Logic and set theory to philosophy. In G. Crocco, E.-M. Engelen, et al., editors, *Gödelian Studies on the Max-Phil Notebooks, Vol. I*. to appear.

[70] Matt Foreman, Menachem Magidor and Saharon Shelah. Martin's maximum, saturated ideals, and nonregular ultrafilters. I. *Ann. of Math. (2)*, 127(1):1–47, 1988.

[71] Abraham A. Fraenkel and Yehoshua Bar-Hillel. *Foundations of Set Theory*. Studies in Logic and the Foundations of Mathematics. North-Holland Publishing Co., Amsterdam, 1958.

[72] Roland Fraïssé. Sur l'extension aux relations de quelques propriétés des ordres. *Ann. Sci. Ecole Norm. Sup. (3)*, 71:363–388, 1954.

[73] Roland Fraïssé. Sur quelques classifications des relations, basées sur des isomorphismes restreints. II. Application aux relations d'ordre, et construction d'exemples montrant que ces classifications sont distinctes. *Publ. Sci. Univ. Alger. Sér. A.*, 2:273–295 (1957), 1955.

[74] Curtis Franks. *The Autonomy of Mathematical Knowledge*. Cambridge University Press, 2009. Hilbert's program revisited.

[75] Greg Frost-Arnold. Was Tarski's theory of truth motivated by physicalism? *Hist. Philos. Logic*, 25(4):265–280, 2004.

[76] Gebhard Fuhrken. A remark on the Härtig quantifier. *Z. Math. Logik Grundlagen Math.*, 18:227–228, 1972.

[77] Robin Gandy. The confluence of ideas in 1936. In *The Universal Turing Machine: A Half-Century Survey*, Oxford Science Publications, pages 55–111. Oxford University Press, New York, 1988.

[78] Michael Glanzberg. Logical consequence and natural language. In *Foundations of Logical Consequence*. Oxford University Press, 2015.

[79] Kurt Gödel. Über formal unentscheidbare Sätze der Principia Mathematica und verwandter Systeme I. *Monatsh. Math. Phys.*, 38(1):173–198, 1931.

[80] Kurt Gödel. The consistency of the axiom of choice and of the generalized continuum hypothesis. *Proc. Natl. Acad. Sci. USA*, 24:556–557, 1938.

[81] Kurt Gödel. *The Consistency of the Continuum Hypothesis.* Annals of Mathematics Studies, no. 3. Princeton University Press, 1940.

[82] Kurt Gödel. What is Cantor's continuum problem? *Amer. Math. Monthly*, 54:515–525, 1947.

[83] Kurt Gödel. Über eine bisher noch nicht benützte Erweiterung des finiten Standpunktes. *Dialectica*, 12:280–287, 1958.

[84] Kurt Gödel. *Collected Works. I: Publications 1929–1936.* (eds. S. Feferman et al.). Oxford University Press, 1986.

[85] Kurt Gödel. *Collected Works. II: Publications 1938–1974.* (eds. S. Feferman et al.). Oxford University Press, 1990.

[86] Kurt Gödel. Remarks before the Princeton Bicentennial Conference of Problems in Mathematics, 1946. In: *Collected Works. Vol. II.* The Clarendon Press Oxford University Press, New York, 1990. Publications 1938–1974, Edited and with a preface by Solomon Feferman.

[87] Kurt Gödel. *Collected Works. III: Unpublished Essays and Lectures.* (eds. S. Feferman et al.). Oxford University Press, 1995.

[88] Kurt Gödel. *Collected Works. IV: Correspondence A-G.* (eds. S. Feferman et al.). Oxford University Press, 2003.

[89] Kurt Gödel. *Collected Works. V: Correspondence H-Z.* (eds. S. Feferman et al.). Oxford University Press, 2003.

[90] Norma B. Goethe and Michèle Friend. Confronting ideals of proof with the ways of proving of the research mathematician. *Stud. Log.*, 96(2):273–288, 2010.

[91] Martin Goldstern and Saharon Shelah. The bounded proper forcing axiom. *J. Symb. Log.*, 60(1):58–73, 1995.

[92] Mario Gómez-Torrente. Tarski on logical consequence. *Notre Dame J. Form. Log.*, 37(1):125–151, 1996.

[93] Mario Gómez-Torrente. Rereading Tarski on logical consequence. *Rev. Symb. Log.*, 2(2):249–297, 2009.

[94] Mario Gómez-Torrente. Alfred tarski. In Edward N. Zalta, editor, *The Stanford Encyclopedia of Philosophy*. Metaphysics Research Lab, Stanford University, spring 2019 edition, 2019.

[95] Ju. Š. Gurevič. The decision problem for standard classes. *J. Symb. Log.*, 41(2):460–464, 1976.

[96] Paul R. Halmos. *Naive Set Theory.* The University Series in Undergraduate Mathematics. D. Van Nostrand Co., Princeton, NJ-Toronto-London-New York, 1960.

[97] Michael Harris. *Mathematics without Apologies: A Portrait of a Problematic Vocation.* Princeton University Press, 2015.

[98] Kai Hauser and W. Hugh Woodin Strong axioms of infinity and the debate about realism. *J. Philos.*, 111:397–419, 2014.

[99] Heinrich Herre, MichałKrynicki, Alexandr Pinus and Jouko Väänänen. The Härtig quantifier: a survey. *J. Symb. Log.*, 56(4):1153–1183, 1991.

[100] Arend Heyting. *Intuitionism: An Introduction.* Second revised edition. North-Holland Publishing Co., Amsterdam, 1966.

[101] D. Hilbert and P. Bernays. *Grundlagen der Mathematik. I.* Zweite Auflage. Die Grundlehren der mathematischen Wissenschaften, Band 40. Springer-Verlag, Berlin-New York, 1968.

[102] David Hilbert. On the infinite. In J. van Heijenoort, editor, *From Frege to Gödel: A Source Book in Mathematical Logic*, pages 367–392. Harvard University Press, Cambridge, MA, 1965.

[103] David Hilbert. Die grundlagen der geometrie. In M. Hallett and U. Majer, editors, *David Hilbert's Lectures on the Foundations of Geometry (1891–1902)*. Springer, 2004.

[104] Wilfrid Hodges. What is a structure theory? *Bull. Lond. Math. Soc.*, 19:209–237, 1987.

[105] Wilfrid Hodges. Detecting the logical content: Burley's "purity of logic". In *We Will Show Them! Essays in Honour of Dov Gabbay on his 60th Birthday, II*. College Publications, King's College London, 2005.

[106] Wilfrid Hodges. Tarski's Theory of Definition. In Douglas Patterson, editor, *New Essays on Tarski and Philosophy*, pages 94–132. Oxford University Press, 2008.

[107] Wilfrid Hodges. Set theory, Model Theory, and Computability theory. In *The Development of Modern Logic*, pages 471–498. Oxford University Press, 2009.

[108] Ehud Hrushovski. The Mordell-Lang conjecture for function fields. *J. Amer. Math. Soc.*, 9(3):667–690, 1996.

[109] Ehud Hrushovski and Boris Zilber. Zariski geometries. *J. Amer. Math. Soc.*, 9(1):1–56, 1996.

[110] Edmund Husserl. *Contributions à la théorie du calcul des variations*, volume 65 of *Queen's Papers in Pure and Applied Mathematics*. Queen's University, Kingston, ON, 1983. Edited and with an introduction by J. Vauthier.

[111] Edmund Husserl. *Ideas Pertaining to a Pure Phenomenology and to a Phenomenological Philosophy*. Kluwer, Dordrecht, 1983. Translated by F. Kersten.

[112] B. Zilber J. A. Cruz Morales and A. Villaveces. Around logical perfection. *Theoria*, to appear.

[113] Thomas Jech. *Set Theory*. Springer Monographs in Mathematics. Springer-Verlag, Berlin, 2003. The third millennium edition, revised and expanded.

[114] Bjarni Jónsson. Homogeneous universal relational systems. *Math. Scand.*, 8:137–142, 1960.

[115] Akihiro Kanamori. Aspect perception and the history of mathematics. preprint, 2014.

[116] Akihiro Kanamori. *The Higher Infinite*. Springer Monographs in Mathematics. Springer-Verlag, Berlin, second edition, 2009. Large cardinals in set theory from their beginnings, Paperback reprint of the 2003 edition.

[117] Carol R. Karp. Finite-quantifier equivalence. In *Theory of Models (Proceedings of the 1963 International Symposium at Berkeley)*, pages 407–412. North-Holland, Amsterdam, 1965.

[118] Matt Kaufmann. Set theory with a filter quantifier. *J. Symb. Log.*, 48(2):263–287, 1983.

[119] H. Jerome Keisler. Models with orderings. In *Logic, Methodology and Philosophy of Science III (Proceedings of the Third International Congress, Amsterdam, 1967)*, pages 35–62. North-Holland, Amsterdam, 1968.

[120] H. Jerome Keisler. Ultraproducts and saturated models. *Nederl. Akad. Wetensch. Proc. Ser. A 67 = Indag. Math.*, 26:178–186, 1964.

[121] H. Jerome Keisler. Logic with the quantifier "there exist uncountably many". *Ann. Math. Logic*, 1:1–93, 1970.

[122] H. Jerome Keisler. *Model Theory for Infinitary Logic. Logic with Countable Conjunctions and Finite Quantifiers*. North-Holland Publishing Co., Amsterdam-London, 1971. Studies in Logic and the Foundations of Mathematics, Vol. 62.

[123] Juliette Kennedy. Turing, Gödel and the "Bright Abyss". In *Philosophical Explorations of the Legacy of Alan Turing*, volume 324 of Boston Studies in Philosophy. Springer.

[124] Juliette Kennedy. On formalism freeness: Implementing Gödel's 1946 Princeton bicentennial lecture. *Bull. Symbolic Logic*, 2013.

[125] Juliette Kennedy. Gödel's 1946 Princeton Bicentennial Lecture: An appreciation. In Juliette Kennedy, editor, *Interpreting Gödel*. Cambridge University Press, 2014.

[126] Juliette Kennedy. On the Logic without Borders point of view. In *Logic without Borders: Essays on Set Theory, Model Theory, Philosophical Logic and Philosophy of Mathematics*, pages 1–14. de Gruyter, 2015.

[127] Juliette Kennedy. Kurt Gödel. In Edward N. Zalta, editor, *The Stanford Encyclopedia of Philosophy*. Metaphysics Research Lab, Stanford University, winter 2018 edition, 2018.

[128] Juliette Kennedy, Menachem Magidor and Jouko Väänänen. Inner models from extended logics: Part 1. https://arxiv.org/pdf/2007.10764.pdf and *Journal of Mathematical Logic, to appear*.

[129] Juliette Kennedy and Jouko Väänänen. Squeezing arguments and strong logics. In *Logic, Methodology and Philosophy of Science: Models and Modelling*, pages 63–83. Coll. Publ., London, 2017.

[130] Jonathan Kirby. On quasiminimal excellent classes. *J. Symb. Log.*, 75(2):551–564, 2010.

[131] S. C. Kleene. A note on recursive functions. *Bull. Amer. Math. Soc.*, 42(8):544–546, 1936.

[132] S. C. Kleene. Recursive predicates and quantifiers. *Trans. Amer. Math. Soc.*, 53:41–73, 1943.

[133] S. C. Kleene. Recursive functionals and quantifiers of finite types. I. *Trans. Amer. Math. Soc.*, 91:1–52, 1959.

[134] S. C. Kleene. Origins of recursive function theory. *Ann. Hist. Comput.*, 3(1):52–67, 1981.

[135] S. C. Kleene. The theory of recursive functions, approaching its centennial. *Bull. Amer. Math. Soc. (N.S.)*, 5(1):43–61, 1981.

[136] S. C. Kleene and J. B. Rosser. The inconsistency of certain formal logics. *Ann. of Math. (2)*, 36(3):630–636, 1935.

[137] Georg Kreisel. Informal rigour and completeness proofs. In *Proceedings of the International Colloquium in the Philosophy of Science, London, 1965, Vol. 1*. Edited by Imre Lakatos, pages 138–157. North-Holland Publishing Co., Amsterdam, 1967.

[138] Georg Kreisel. Which number theoretic problems can be solved in recursive progressions on Π_1^1-paths through O? *J. Symb. Log.*, 37:311–334, 1972.

[139] Georg Kreisel. Logical Foundations: A lingering malaise. *Philosophisches Archiv der Universität Konstanz*, 1984.

[140] Georg Kreisel. Second thoughts around some of Gödel's writings: a non-academic option. *Synthese*, 114: 99–160, 1998. Gödel, I.

[141] Georg Kreisel. Informal rigour and completeness proofs. *Studies in Logic and the Foundations of Mathematics*, volume 47, pages 138–186, 1967.

[142] Georg Kreisel. Mathematical logic: tool and object lesson for science. *Synthese*, 62:139–151, 1985. The present state of the problem of the foundations of mathematics (Florence, 1981).

[143] Goerg Kreisel. Mathematics without foundations. *J. Symbolic Logic*, 37(2):402–404, 1972.

[144] Saul Kripke. The Church-Turing "Thesis" as a Special Corollary of Gödel's Completeness Theorem. In B. Jack Copeland, Carl J. Posy, and Oron Shagrir, editors, *Computability: Gödel, Church, and Beyond*. MIT Press, Cambridge, 2013.

[145] David W. Kueker. Abstract elementary classes and infinitary logics. *Ann. Pure Appl. Logic*, 156(2-3):274–286, 2008.

[146] Kenneth Kunen. *Set Theory*, volume 34 of Studies in Logic (London). College Publications, London, 2011.

[147] Brendan Larvor. How to think about informal proofs. *Synthese*, 187(2):715–730, 2012.

[148] Sandra Laugier. *Why We Need Ordinary Language Philosophy*. University of Chicago Press, 2013.

[149] Dwight E. Lee and Robert N. Beck. The meaning of "historicism". *Am. Hist. Rev.*, 59:568–577, 1954.

[150] Stanislaw Leśniewski. Fundamentals of a new system of the foundations of mathematics. In *Collected Works* Dordrecht: Kluwer Academic Publishers,1992.

[151] Azriel Lévy. Axiom schemata of strong infinity in axiomatic set theory. *Pacific J. Math.*, 10:223–238, 1960.

[152] Per Lindström. First order predicate logic with generalized quantifiers. *Theoria*, 32:186–195, 1966.

[153] Per Lindström. On extensions of elementary logic. *Theoria*, 35:1–11, 1969.

[154] Per Lindström. A characterization of elementary logic. In *Modality, Morality and Other Problems of Sense and Nonsense*, pages 189–191. Lund, Gleerup, 1973.

[155] Nikolai Luzin. Sur les ensembles projectifs de m. Henri Lebesgue. *C. R. Acad. Sci. Paris*, 180(2):1572–1574, 1925.

[156] Angus Macintyre. Ramsey quantifiers in arithmetic. In *Model Theory of Algebra and Arithmetic (Proc. Conf., Karpacz, 1979)*, volume 834 of Lecture Notes in Mathematics, pages 186–210. Springer, Berlin, 1980.

[157] Saunders MacLane. Book review: *The Logical Syntax of Language*. *Bull. Amer. Math. Soc.*, 44(3):171–176, 1938.

[158] Saunders MacLane. *Mathematics: Form and Function*. Springer, 1986.

[159] Penelope Maddy. *Second Philosophy*. Oxford University Press, 2009. A naturalistic method.

[160] Penelope Maddy. *Defending the Axioms: On the Philosophical Foundations of Set Theory*. Oxford University Press, 2011.

[161] Penelope Maddy. Some naturalistic reflections on set theoretic method. *Topoi*, 20(1):17–27, 2001.

[162] Menachem Magidor and Jouko Väänänen. On Löwenheim-Skolem-Tarski numbers for extensions of first order logic. *J. Math. Log.*, 11(1):87–113, 2011.

[163] Ofra Magidor. The last dogma of type confusions. In *Proc. Arist. Soc.*, 109:1–29, 2009.

[164] Ofra Magidor. *Category Mistakes*. Oxford University Press, 2013.

[165] Johann Makowsky, Saharon Shelah, and Jonathan Stavi. Δ-logics and generalized quantifiers. *Ann. Math. Logic*, 10(2):155–192, 1976.

[166] Paolo Mancosu. *From Brouwer to Hilbert*. Oxford University Press, New York, 1998. The debate on the foundations of mathematics in the 1920s, With the collaboration of Walter P. van Stigt, Reproduced historical papers translated from the Dutch, French and German.

[167] Paolo Mancosu. Harvard 1940-1941: Tarski, carnap and quine on a finitistic language of mathematics for science. *Hist. Phil. Log.*, 26(4):327–357, 2005.

[168] Paolo Mancosu. Tarski on models and logical consequence. In *The Architecture of Modern Mathematics*, pages 209–237. Oxford University Press, 2006.

[169] Paolo Mancosu. Tarski, Neurath and Kokoszyńska on the semantic conception of truth. In Douglas Patterson, editor, *New Essays on Tarski and Philosophy*, pages 192–224. Oxford University Press, 2008.

[170] Yuri Manin. *A Course in Mathematical Logic for Mathematicians*, volume 53 of Graduate Texts in Mathematics. Springer, New York, second edition, 2010. Chapters I–VIII translated from the Russian by Neal Koblitz, With new chapters by Boris Zilber and the author.

[171] Donald A. Martin. Borel determinacy. *Ann. of Math. (2)*, 102(2):363–371, 1975.

[172] Donald A. Martin. Completeness or incompleteness of basic mathematical concepts. In Peter Koellner, editor, *Exploring the Frontiers of Infinity*. 2012.

[173] Donald A. Martin and John R. Steel. Projective determinacy. *Proc. Nat. Acad. Sci. U.S.A.*, 85(18):6582–6586, 1988.

[174] Kenneth McAloon. Consistency results about ordinal definability. *Ann. Math. Logic*, 2(4):449–467, 1970/71.

[175] Vann McGee. Logical operations. *J. Philos. Logic*, 25(6):567–580, 1996.

[176] Colin McLarty. Poincaré: mathematics & logic & intuition. *Philos. Math. (3)*, 5(2):97–115, 1997.

[177] Friederike Moltmann. Reference to numbers in natural language. *Philos. Stud.*, 162(3):499–536, 2013.

[178] Richard Montague. Reduction of higher-order logic. In *Theory of Models (Proceedings of the 1963 International Symposium at Berkeley)*, pages 251–264. North-Holland, Amsterdam, 1965.

[179] Justin Tatch Moore. Set mapping reflection. *J. Math. Log.*, 5(1):87–97, 2005.

[180] Andrzej Mostowski. On a generalization of quantifiers. *Fund. Math.*, 44:12–36, 1957.

[181] Andrzej Mostowski. *Thirty Years of Foundational Studies. Lectures on the Development of Mathematical Logic and the Study of the Foundations of Mathematics in 1930–1964*. Acta Philosophica Fennica, Fasc. XVII. Barnes & Noble, Inc., New York, 1966.

[182] Andrzej Mostowski. Alfred Tarski. In *The Encyclopaedia of Philosophy,*, volume 8, pages 77–81. Macmillan, 1967.

[183] Andrzej Mostowski. *Foundational Studies. Selected Works. Vol. I*, volume 93 of Studies in Logic and the Foundations of Mathematics. North-Holland Publishing Co., Amsterdam-New York; PWN – Polish Scientific Publishers, Warsaw, 1979. Edited by Kazimierz Kuratowski, Wiktor Marek, Leszek Pacholski, Helena Rasiowa, Czesław Ryll-Nardzewski and PawełZbierski, With contributions by Wiktor Marek, A. Grzegorczyk, W. Guzicki, Leszek Pacholski, C. Rauszer and PawełZbierski.

[184] Daniele Mundici and Wilfried Sieg. Computability theory. *Report CMU-PHIL-54*, 1994.

[185] Roman Murawski and Jan Woleński. Tarski and his Polish predecessors on truth. In Douglas Patterson, editor, *New Essays on Tarski and Philosophy*, pages 21–43. Oxford University Press, 2008.

[186] Roman Murawski and JanWoleński. Tarski and his Polish predecessors on truth. In Douglas Patterson, editor, *New Essays on Tarski and Philosophy*, pages 72–93. Oxford University Press, 2008.

[187] John Myhill and Dana Scott. Ordinal definability. In *Axiomatic Set Theory (Proceedings of Symposia in Pure Mathematics, Vol. XIII, Part I, Univ. California, Los Angeles, Calif., (1967)*, pages 271–278. American Mathematical Society, Providence, R.I., 1971.

[188] Mark E. Nadel. An arbitrary equivalence relation as elementary equivalence in an abstract logic. *Z. Math. Logik Grundlag. Math.*, 26(2):103–109, 1980.

[189] Douglas Patterson. Tarski's conception of meaning. In Douglas Patterson, editor, *New Essays on Tarski and Philosophy*, pages 72–93. Oxford University Press, 2008.

[190] Rózsa Péter. Konstruktion nichtrekursiver Funktionen. *Math. Ann.*, 111(1):42–60, 1935.

[191] Rózsa Péter. Über die mehrfache Rekursion. *Math. Ann.*, 113(1):489–527, 1937.

[192] Parsons, Charles. Reference, Rationality, and Phenomonology: Themes from Follesdal. In Michael Frauchiger, editor, *Some consequences of the entanglement of logic and mathematics*. Ontos Verlag, Frankfurt, 2013.

[193] Anand Pillay and Charles Steinhorn. Definable sets in ordered structures. *Bull. Amer. Math. Soc. (N.S.)*, 11(1):159–162, 1984.

[194] Henri Poincaré. Analysis situs. *Journal de l'École Polytechnique*, 2(1):1–123, 1895.

[195] Emil L. Post. Finite combinatory processes-formulation 1. *J. Symb. Log.*, 1:103–105, 1936.

[196] Emil L. Post. Recursively enumerable sets of positive integers and their decision problems. *Bull. Amer. Math. Soc.*, 50:284–316, 1944.

[197] Emil L. Post. *Solvability, Provability, Definability: The Collected Works of Emil L. Post*. Contemporary Mathematicians. Birkhäuser Boston, Inc., Boston, MA, 1994. Edited and with an introduction by Martin Davis.

[198] Pavel Pudlák. On the lengths of proofs of consistency: a survey of results. In *Collegium Logicum, Vol. 2*, pages 65–86. Springer, Vienna, 1996.

[199] Hilary Putnam. Mathematics without foundations. *J. Philos.*, 64(1):5–22, 1967.

[200] Hilary Putnam. Models and reality. *J. Symb. Log.*, 45(3):464–482, 1980.

[201] Hilary Putnam. *The Collapse of the Fact/Value Dichotomy and Other Essays.* Harvard University Press, Cambridge, MA, 2004.

[202] Hilary Putnam. *Ethics without Ontology.* Harvard University Press, Cambridge, MA, 2004.

[203] Willard Van Orman Quine. Carnap and logical truth. In *The Ways of Paradox and Other Essays*, pages 107–132. Harvard University Press, Cambridge, MA, 1976.

[204] Willard Van Orman Quine. *Philosophy of Logic.* Harvard University Press, Cambridge, MA, second edition, 1986.

[205] Yehuda Rav. A critique of a formalist-mechanist version of the justification of arguments in mathematicians' proof practices. *Philos. Math. (3)*, 15(3):291–320, 2007.

[206] Michael Rescorla. Church's thesis and the conceptual analysis of computability. *Notre Dame J. Formal Logic*, 48(2):253–280 (electronic), 2007.

[207] Abraham Robinson. *On the Metamathematics of Algebra.* Studies in Logic and the Foundations of Mathematics. North-Holland Publishing Co., Amsterdam, 1951.

[208] Abraham Robinson. Metamathematical problems. *J. Symb. Log.*, 38:500–516, 1973.

[209] Francisco Rodríguez-Consuegra. Tarski's intuitive notion of set. In *Essays on the Foundations of Mathematics and Logic, Volume 1*, pages 227–266. Polimetrica s.a.s., 2005.

[210] Michael Rohlf. The ideas of pure reason. In *The Cambridge Companion to Kant's Critique of Pure Reason*, pages 190–210. Cambridge University Press, 2010.

[211] Bertrand Russell. *Mysticism and Logic.* Dover Publications, 1917.

[212] Gerald E. Sacks. *Saturated Model Theory.* World Scientific Publishing Co. Pte. Ltd., Hackensack, NJ, second edition, 2010.

[213] Gil Sagi. Logicality and meaning. *Rev. Symb. Log.*, 11(1):133–159, 2018.

[214] James H. Schmerl and Stephen G. Simpson. On the role of Ramsey quantifiers in first order arithmetic. *J. Symb. Log.*, 47(2):423–435, 1982.

[215] Dana Scott. Logic with denumerably long formulas and finite strings of quantifiers. In *Theory of Models (Proceedings of the 1963 International Symposium at Berkeley)*, pages 329–341. North-Holland, Amsterdam, 1965.

[216] Stewart Shapiro. Acceptable notation. *Notre Dame J. Formal Logic*, 23(1):14–20, 1982.

[217] Stewart Shapiro. Review: Stephen C. Kleene, *Origins of Recursive Function Theory*; Martin Davis, *Why Godel Didn't Have Church's Thesis*; Stephen C. Kleene, *Reflections on Church's Thesis. J. Symbolic Logic*, 55(1):348–350, 03 1990.

[218] Stewart Shapiro. *Foundations without Foundationalism*, volume 17 of *Oxford Logic Guides*. The Clarendon Press, Oxford University Press, New York, 1991. A case for second-order logic, Oxford Science Publications.

[219] Stewart Shapiro. The open texture of computability. In *Computability – Turing, Gödel, Church, and beyond*, pages 153–181. MIT Press, Cambridge, MA, 2013.

[220] Saharon Shelah. Every two elementarily equivalent models have isomorphic ultrapowers. *Israel J. Math.*, 10:224–233, 1971.

[221] Saharon Shelah. Infinite abelian groups, Whitehead problem and some constructions. *Israel J. Math.*, 18:243–256, 1974.

[222] Saharon Shelah. Generalized quantifiers and compact logic. *Trans. Amer. Math. Soc.*, 204:342–364, 1975.

[223] Saharon Shelah. *Classification Theory and the Number of Nonisomorphic Models*, volume 92 of Studies in Logic and the Foundations of Mathematics. North-Holland Publishing Co., Amsterdam-New York, 1978.

[224] Saharon Shelah. Classification theory for nonelementary classes. I. The number of uncountable models of $\psi \in L_{\omega_1,\omega}$. Part A. *Israel J. Math.*, 46(3):212–240, 1983.

[225] Saharon Shelah. On the no(M) for M of singular power. In *Around Classification Theory of Models*, volume 1182 of Lecture Notes in Mathematics, pages 120–134. Springer, Berlin, 1986.

[226] Saharon Shelah. *Classification Theory for Abstract Elementary Classes*, volume 18 of Studies in Logic: Mathematical logic and foundations. College Publications, 2009.

[227] Saharon Shelah. *Classification Theory for Abstract Elementary Classes 2* , volume 20 of Studies in Logic: Mathematical logic and foundations. College Publications, 2009.

[228] Saharon Shelah. Nice infinitary logics. *J. Amer. Math. Soc.*, 25(2):395–427, 2012.

[229] Saharon Shelah. Introduction. In *Beyond First Order Model Theory*, pages 131–187. CRC Press, Boca Raton, FL, 2017.

[230] Gila Sher. *The Bounds of Logic*. A Bradford Book. MIT Press, Cambridge, MA, 1991. A generalized viewpoint.

[231] Gila Sher. Tarski's thesis. In Douglas Patterson, editor, *New Essays on Tarski and Philosophy*, pages 300–339. Oxford University Press, 2008.

[232] J. R. Shoenfield. The problem of predicativity. In *Essays on the Foundations of Mathematics*, pages 132–139. Magnes Press, Hebrew University, Jerusalem, 1961.

[233] Wilfried Sieg. Step by recursive step: Church's analysis of effective calculability. *Bull. Symbolic Logic*, 3(2):154–180, 1997.

[234] Wilfried Sieg. Only two letters: the correspondence between Herbrand and Gödel. *Bull. Symbolic Logic*, 11(2):172–184, 2005.

[235] Wilfried Sieg. Gödel on computability. *Philos. Math.*, 14:189–207, 2006.

[236] Wilfried Sieg. On computability. In *Philosophy of mathematics, Handbook of the Philosophy of Science*, pages 535–630. Elsevier/North-Holland, Amsterdam, 2009.

[237] Hourya Benis Sinaceur. Address at the Princeton University Bicentennial Conference on problems of mathematics (December 17-19, 1946), by Alfred Tarski. *Bull. Symbolic Logic*, 6(1):1–44, 2000.

[238] Hourya Benis Sinaceur. Tarski's practice and philosophy: Between formalism and pragmatism. In Lindström, Palmgren, Segerberg, and Stoltenberg-Hansen, editors, *Logicism, Intuitionism, and Formalism, What Has Become of Them?*, volume 341 of Synthese Library Springer, 2009.

[239] Timothy Smiley. Aristotle's completeness proof. *Anc. Philos.*, 14:25–38, 1994.

[240] Peter Smith. Squeezing arguments. *Analysis (Oxford)*, 71(1):23–30, 2011.

[241] Robert Soare. Why Turing's Thesis is not a Thesis. In Thomas Stram and Giovanni Sommaruga, editors, *Turing Centenary Volume*. Birkhäuser/Springer, 2015.

[242] Robert Soare. Computability and recursion. *Bull. Symbolic Logic*, 2(3):284–321, 1996.

[243] Jonathan Stavi and Jouko Väänänen. Reflection principles for the continuum. In *Logic and Algebra*, volume 302 of Contemporary Mathematics, pages 59–84. American Mathematical Society, Providence, RI, 2002.

[244] Martin Stockhof. Can natural language be captured in a formal system? In *Handbook of Formal Philosophy*. Springer, forthcoming.

[245] Alfred Tarski. Sur les ensembles définissables de nombres réels. *Fund. Math.*, (7):210–39, 1931.

[246] Alfred Tarski. Der Wahrheitsbegriff in den formalisierten Sprachen. Stud. Philos. 1, 261–405 (1935)., 1935.

[247] Alfred Tarski. The semantic conception of truth and the foundations of semantics. *Philos. and Phenomenol. Res.*, 4:341–376, 1944.

[248] Alfred Tarski. *A Decision Method for Elementary Algebra and Geometry*. RAND Corporation, Santa Monica, CA, 1948.

[249] Alfred Tarski. Some notions and methods on the borderline of algebra and metamathematics. In *Proceedings of the International Congress of Mathematicians, Cambridge, Mass., 1950, vol. 1*, pages 705–720. American Mathematical Society, Providence, RI, 1952.

[250] Alfred Tarski. Contributions to the theory of models. III. *Nederl. Akad. Wetensch. Proc. Ser. A.*, 58:56–64 = Indagationes Math. 17, 56–64 (1955), 1955.

[251] Alfred Tarski. *Logic, Semantics, Metamathematics*. Hackett Publishing Co., Indianapolis, IN, second edition, 1983. Papers from 1923 to 1938, Translated by J. H. Woodger, Edited and with an introduction by John Corcoran.

[252] Alfred Tarski. What are logical notions? *Hist. Philos. Logic*, 7(2):143–154, 1986. Edited by John Corcoran.

[253] Alfred Tarski. *Introduction to Logic and to the Methodology of Deductive Sciences*. Dover Publications, Inc., New York, 1995. Translated from the 1936 Polish original by Olaf Helmer, reprint of the 1946 translation.

[254] Alfred Tarski and Robert L. Vaught. Arithmetical extensions of relational systems. *Compositio Math*, 13:81–102, 1958.

[255] Leslie H. Tharp. The characterization of monadic logic. *J. Symb. Log.*, 38:481–488, 1973.

[256] Charles Travis. Insensitive semantics. *Mind Lang.*, (3):39–49, 2006.

[257] S. Ulam. Zur Maßtheorie in der allgemeinen Mengenlehre. *Fundam. Math.*, 16:140–150, 1930.

[258] Alasdair Urquhart. Emil Post. In *Handbook of the History of Logic. Vol. 5. Logic from Russell to Church*, pages 617–666. Elsevier/North-Holland, Amsterdam, 2009.

[259] Jouko Väänänen. Set-theoretic definability of logics. In *Model-Theoretic Logics*, Perspectives in Logic, pages 599–643. Springer, New York, 1985.

[260] Jouko Väänänen. Two axioms of set theory with applications to logic. *Ann. Acad. Sci. Fenn. Ser. A I Math. Dissertationes*, (20):19, 1978.

[261] Jouko Väänänen. Abstract logic and set theory. I. Definability. In *Logic Collo-quium '78 (Mons, 1978)*, volume 97 of Studies in Logic and the Foundations of Mathematics, pages 391–421. North-Holland, Amsterdam-New York, 1979.

[262] Jouko Väänänen. Boolean-valued models and generalized quantifiers. *Ann. Math. Logic*, 18(3):193–225, 1980.

[263] Jouko Väänänen. Abstract logic and set theory. II. Large cardinals. *J. Symb. Log.*, 47(2):335–346, 1982.

[264] Jouko Väänänen. Second-order logic and foundations of mathematics. *Bull. Symbolic Logic*, 7(4):504–520, 2001.

[265] Jouko Väänänen. *Models and Games*, volume 132 of Cambridge Studies in Advanced Mathematics. Cambridge University Press, 2011.

[266] Jouko Väänänen. Second order logic or set theory? *Bull. Symbolic Logic*, 18(1):91–121, 2012.

[267] Jouko Väänänen. Second-order and higher-order logic. In Edward N. Zalta, editor, *The Stanford Encyclopedia of Philosophy*. Metaphysics Research Lab, Stanford University, fall 2019 edition, 2019.

[268] Mark van Atten and Juliette Kennedy. On the philosophical development of Kurt Gödel. *Bull. Symbolic Logic*, 9(4):425–476, 2003.

[269] Mark van Atten and Juliette Kennedy. Gödel's modernism: On set-theoretic incompleteness, revisited. In Lindström, Palmgren, Segerberg, and Stoltenberg-Hansen, editors, *Logicism, Intuitionism, and Formalism, What Has Become of Them?*, volume 341 of Synthese Library Springer, 2009.

[270] Lou van den Dries. A generalization of the Tarski-Seidenberg theorem, and some nondefinability results. *Bull. Amer. Math. Soc. (N.S.)*, 15(2):189–193, 1986.

[271] B. L. van der Waerden. *Modern Algebra. Vol. I*. Frederick Ungar Publishing Co., New York, NY, 1949. Translated from the second revised German edition by Fred Blum, with revisions and additions by the author.

[272] Robert L. Vaught. Denumerable models of complete theories. In *Infinitis-tic Methods (Proceedings of the Symposium on Foundations of Mathematics, Warsaw, 1959)*, pages 303–321. Pergamon, Oxford; Państwowe Wydawnictwo Naukowe, Warsaw, 1961.

[273] Robert L. Vaught. Alfred Tarski's work in model theory. *J. Symb. Log.*, 51(4):869–882, 1986.

[274] Robert L. Vaught. Errata: "Alfred Tarski's work in model theory". *J. Symb. Log.*, 52(4):vii, 1987.

[275] Oswald Veblen and John Wesley Young. *Projective Geometry*. Vol. I, II. New York-Toronto-London: Blaisdell Publishing Company. 1965.

[276] Petr Vopenka, Bohuslav Balcar and Petr Hajek. The notion of effective sets and a new proof of the consistency of the axiom of choice (abstract). 1968.

[277] Walter Biemel, editor. *Phänomenologische Psychologie: Vorlesungen Sommersemester 1925*. Nijhoff, Den Haag, 1962.

[278] Hao Wang. *From Mathematics to Philosophy*. Routledge, 1974.

[279] Hao Wang. Some facts about Kurt Gödel. *J. Symb. Log.*, 46(3):653–659, 1981.

[280] Hao Wang. *A Logical Journey*. MIT Press, Cambridge, MA, 1996.

[281] Martin Weese. Generalized Ehrenfeucht games. *Fund. Math.*, 109(2):103–112, 1980.

[282] Philip D. Welch. The ramified analytical hierarchy using extended logics. *Bull. Symb. Log.*, 24(3):306–318, 2018.

[283] Alex James Wilkie. Model completeness results for expansions of the ordered field of real numbers by restricted Pfaffian functions and the exponential function. *J. Amer. Math. Soc.*, 9(4):1051–1094, 1996.

[284] Mark Wilson. *Wandering Significance: An Essay on Conceptual Behavior.* Oxford University Press, 2008.

[285] Mark Wilson. *Physics Avoidance.* Oxford University Press, 2017.

[286] William Wimsatt. The ontology of complex systems: Levels of organization, perspectives, and causal thickets. *Can. J. Philos.*, 20:207–274, 1994.

[287] W. Hugh Woodin. Supercompact cardinals, sets of reals, and weakly homogeneous trees. *Proc. Nat. Acad. Sci. U.S.A.*, 85(18):6587–6591, 1988.

[288] W. Hugh Woodin. *The Axiom of Determinacy, Forcing Axioms, and the Nonstationary Ideal*, volume 1 of De Gruyter Series in Logic and its Applications. Walter de Gruyter GmbH & Co. KG, Berlin, revised edition, 2010.

[289] Boris Zilber. Pseudo-exponentiation on algebraically closed fields of characteristic zero. *Ann. Pure Appl. Logic*, 132(1):67–95, 2005.

[290] Boris Zilber. A categoricity theorem for quasi-minimal excellent classes. In *Logic and Its Applications*, volume 380 of Contemporay Mathematics, pages 297–306. American Mathematical Society, Providence, RI, 2005.

Index

Printed in the United States
by Baker & Taylor Publisher Services

Printed in the United States
by Baker & Taylor Publisher Services